a table | in the desert

MAKING SPACE HOLY

a table | in the desert

MAKING SPACE HOLY

W. PAUL JONES

PARACLETE PRESS
BREWSTER, MASSACHUSETTS

Library of Congress Cataloging-in-Publication Data

Jones, W. Paul (William Paul)
    A table in the desert : making space holy / W. Paul Jones.
        p. cm.
    ISBN 1-55725-270-X
1. Spiritual life—Catholic Church. 2. Space and time—Religious aspects—Catholic Church. I. Title.
    BX2350.65 .J665 2000
    263'.042—dc21   00-012093

10 9 8 7 6 5 4 3 2 1

Published by Paraclete Press
Brewster, Massachusetts
www.paracletepress.com

Printed in the United States of America.

To

The Roman Catholic Church,
the home of my latter-day pilgrimage,

And To

The United Methodist Church,
the Protestant Church of my upbringing and middle years.

Just As

the marks of ordination by both Churches
have branded my soul for life,

So

may these traditions come to know how much
they are intertwined—for eternity.

In

gratefulness for having been called
to be a Liaison between them.

"They spoke against God, saying, 'Can God spread a table in the desert?'" (Ps. 78:19)

The psalmist felt called to "utter dark sayings from of old," "heard and known" from their own parents. No longer dare we "hide them," even from "the coming generation." Instead, we must tell as our own "the glorious deeds of the Lord." (Ps. 78:2-4)

A strange God it was who called us forth from the land of Egypt. "He divided the sea and let [us] pass through it." By day God led us with a cloud and by night a "fiery light." He led us into the desert, and there he wooed us as God's own fair maiden. And when we thirsted, tenderly God "made streams come out of the rock." (Ps. 78:13-16)

Yet we rebelled against "the Most High in the desert." We "tested God in [our] heart by demanding the food [we] craved." "Can [God] also give bread?" (Ps. 78:17-18, 20)

So God gave us manna, the food of angels. (Ps. 78:24-25)

But still we rebelled, taunting: "Can God spread a table in the desert?" (Ps. 78:19)

—

Each of us must answer these questions:
Is it true that "even though I walk through the valley of the shadow of death," "thou preparest a table before me in the presence of my enemies..."? And on it, "my cup overflows"? (Ps. 23:4, 5 KJV)

—

God told Moses of the Tabernacle he should build. Within it "you shall make a table," and "you shall set the bread of the Presence on the table before me always." (Exod. 25:23, 30)

—⊢—

They came into a large upper room. There he gave them to eat and to drink, promising that those "who have stood by me in my trials" shall "eat and drink at my table in my kingdom." (Luke 22:28, 30)

—⊢—

Even now we have a Great High Priest, that "through the curtain" of his flesh we may enter the heavenly Holy of Holies, and "enter the sanctuary by [his] blood." (Heb. 10:20, 19) Surrounded "by so great a cloud of witnesses" (12:1), "through him, then, let us continually offer a sacrifice of praise to God." (13:15)

—⊢—

Standing before the Table, prepared in our desert, is the One who is "the beginning and the end." (Rev. 21:6)

—⊢—

"Let everyone who is thirsty come [to the table]. Let anyone who wishes take the water of life as a gift." (Rev. 22:17)

——⊢——

The spatial dimension is no less decisive than the temporal in the concrete accomplishment of the mystery of the Incarnation. My meditation turns to the "places" in which God has chosen to "pitch his tent" among us. (Jn.1:14; Ex. 40:34-35; 1 Kngs. 8:10-13) God is equally present in every corner of the earth, so that the whole world may be considered the "temple" of God's presence. Yet this does not take away from the fact that, just as time can be marked by *kairoi*, by special movements of grace, space too may by analogy bear the stamp of particular saving actions of God. This is an intuition present in all religions, that sacred times and sacred spaces are where the encounter with the divine may be experienced more intensely than it would normally be in the vastness of the cosmos.

—Statements from the letter "Concerning Pilgrimages ..."
by Pope John Paul II, June 29, 1999

PRELUDE

What do I do now?
I listen to water
Falling
Into
the
Gentleness
Of being
Nothing
More
Than liquid sound.
And I, at last,
Want nothing
More.

—*W. Paul Jones*

CONTENTS

chapter six
THOUGHTS ON GOD 225

chapter seven
CONCLUSION 255

appendices 261

endnotes 287

1

AN OVERTURE

Some months ago, my book A Season in the Desert: Making Time Holy *was published. This present book on space serves as volume two in the set, completing the Christian exploration of spirituality as "sacralization." Life, by its very nature, is an event within space-time. Christian spirituality, thus, entails living intentionally within that spatiotemporal arena as it is transformed by those events in space and time which the Christian heralds as revelation. These two volumes have been written so as to stand independently, and yet, in another sense, they form a whole.*

1. Looking Back

In the first volume, we acknowledged that when meeting persons it is helpful to get to know them a bit before sharing together important thoughts and ideas. Particularly is this so in exploring space, for there is no escaping the "place" of one's pilgrimage. I was born in Appalachia, where even the anatomy of the land was for me vitally formative. I was amazed at seeing the Rocky Mountains for the first time. It was more than their size. As one sensitive to spirituality, I could tell at a glance that one climbs those mountains in order to return. One goes *up* in order to come *down*. One climbs them for the vista, the awe, the sweep, the whole—but one does not live there. In the Appalachian Mountains, actually hills by comparison, the roads avoid going over them. We travel the ridge, so that one goes *down* in order to come *up*. Propped up on the side of these hills, the coal-mining town of my birth formed me. The mines are deep shafts sliced into the earth. We went in, in order to come out. Many never came out. Even for those of us who did, that space formed us for a lifetime.

I understand the power, dare I say the meaning, of those mysterious drawings on the walls of the caves in southern France and elsewhere. In a deep sense, it has something to do with "going home"—where the maternal womb from which each of us emerged finds a haunting familiarity within the silent wombs of Mother Earth's own emergence. Like touches like. In the mines, the wind drawn through the shafts never stops, like a maternal breathing. Everything is wet, as was the sack of our birthing. The sides glisten in the flickering flame of the carbide lanterns strapped to our foreheads—our own phylacteries. To go in, miles in, is to go back through the space of one's own birthing to the nothingness that preceded it—for death and nothingness mark the shadows and the crevices. The low, backbreaking ceiling is propped with locust poles brought in from the hills. But the roof never stops its cracking sounds, as the earth heaves incessantly. A small bird, more sensitive than we to the smells that are fatal: that is the thin margin of error on which we relied. The carbide could explode, the roof collapse, the way out become blocked. This is all about *space*. I do not know anyone who was not maimed by it—or never came out.

In a powerful sense, spirituality for all of us is this arduous pilgrimage to find home as "sacred space." "While they dance they will sing: 'In you all find their home'" (Ps. 87:7 GRAIL). For many of us in that Appalachian town, this struggle was especially hard in our early years, with each of us living a gray sameness in a "company house"—identical with every other, in rows mounting the hill from the mouths of the mines. Alcohol was the "internal home" for many found at the bar just outside the gate. For women, working hard even before sunrise, the tenor of life had so much to do with waiting. But in each of us is a yearning that is deep: someday to own one's own special place, to drive one's very own pickup; it's all about *space*. And if one is deprived of such space in which to enclose one's "creating," one is still driven—this time to leave one's destructive marks on any space, including one's own self. One of the great Russian Orthodox theologians of our time, Nicolas Berdyaev, distilled his whole lifetime of theological writings in three short sentences. I recall that they went something like this: "Some people speak of this earth as though it was their home. I could never say that. My home does not yet exist, but has to be created—as the Kingdom promised yet to be."

I resonate with him. Though coal dust remains in my lungs, in a deeper sense the hills of my home are not finally home. Years later, I contemplated whether I should be buried there on the hill overlooking the sulfur creek, in a spot waiting for me beside my parents. I remember thinking, if only as the hearse passed by there would be someone playing Mozart through an open window, then I could go home. No such possibility. You can't go home. Home has to do with sacred space, yet to be created. The God in whose image we are created is that of the Creator God— the One who fashioned humans after flinging stars all over the Milky Way, fashioning the planets by name, and cupping with proud hands the waters of the sea. We have been fashioned *in the image of that Creator God*, called to be *co-creators with God*. So it was in Eden that our very reason for being was declared to be as the gardeners of Creation. The Christian's prayer, from the words of Jesus himself, makes the focus clear: "Thy Kingdom come, *on earth....*" It is not enough simply to declare that creating is one of God's pastimes. God *is* Creator, intensely involved in creating to the very center of God's own Being.

The Triune God is indeed Redeemer and Inspirer as well, but in both cases it is *creation* that is being redeemed and inspired. As Athanasius declared, "When you took our lowly nature you transformed our sinful world." Jesus speaks in deep imagery when he insists that in clothing the naked and feeding the hungry we are doing it to God: "'For I was hungry and you gave me food, I was thirsty and you gave me something to drink, I was a stranger and you welcomed me, I was naked and you gave me clothing, I was sick and you took care of me, I was in prison and you visited me'" (Matt. 25:35-36). It all has to do with *space*—with things, with matter, with creation, and with the call to create beauty.

A spiritual breakthrough occurs for the Christian, then, when one recognizes that our internal passion for sacred space is of a piece with the yearnings of the cosmos itself. The Christian faith is intensely spatial and thus unique. Many other religions are intent on the nonspatial, the nonmaterial—on the spiritual as unclothed from the fleshly. Even traditional theism affirms a "God" who is without breadth or height or width or length—for God is a "no-thing." In contrast, the "scandal" of Christianity is its materialism, its concreteness, *its preoccupation with space*. The heart of the gospel is the declaration that the spaceless God has entered space, and the nonmaterial God has taken on flesh. This is not a general sort of suggestion or metaphor. It is an unabashed declaration concerning a *particular* space. Yet the Christian disclosure about that event in a thirty-mile diameter of the "near-East" is not even restricted to a thirty-three-year interval. The resurrection is not a setting aside of this one divine-human fling, signifying a return of an uncloaked Spirit to some divine domain from whence it came. Flesh and blood are integrally involved.

As a theologian-poet I have been grasped by the significance of the "Ascension" and its imagery—that Jesus of Nazareth "sits on the throne." Christ, within the Trinity as God, remains the one with wounded hands and feet, bloody brow, and speared rib cage. Thus the Christ-event is the disclosure that the flesh of our space-time is being taken up into the becoming of God. Another way of putting this is by remembering that Jesus declared that the

Kingdom of God is the new heaven and the new *earth*. His promise has to do with tears and death and thirst, all swallowed up within the sacred space of "the holy city Jerusalem." "It has the glory of God, and a radiance like a very rare jewel, like jasper, clear as crystal." The writer of Revelation strains over these things that compose its space—its "length and width and height," its gates and walls: gold, adorned with jasper, sapphire, agate, emerald, onyx, carnelian, chrysolite, beryl, topaz, chrysoprase, jacinth, amethyst (Rev. 21:11-20). In some mysterious way, it all has to do with space.

Little did I know, those many years ago, as I emerged from the space of the mine's darkness into the growing twilight of my tortured hills, that such a vision was implicit within this little Appalachian space of my first love. How could I have known then that such a little space counted as foretaste in the huge space of God's doing? How could I have known that truly to go down is to come up, to go in is to come out, and by traveling the ridge one can discover a total way of life? Yet to explore such connections has become my life's work.

In her own way, this has been the work of the Church as well. Over forty years ago, I was ordained a United Methodist minister—"called" within that tiny church on Main Street, across from the movie house. I knew what that meant, and over the years I nurtured the art of being a *preacher* and *pastor* and *teacher*. Yet there has been a nagging sense that there was more. Yearning for this "more" led me not long ago to ordination as a Roman

Catholic priest. Last Sunday I was such a priest, standing at the altar, receiving the works and dreams and tears and suffering of all people everywhere in the tokens of bread and wine. "This is my body which will be given up for you." Consecrating them into the ongoing sacrifice of Christ, I elevate them as offering into God. "Through him, with him, in him, in the unity of the Holy Spirit, all glory and honor is yours, almighty Father, for ever and ever." And all of Creation responds, "AMEN!"

The Incarnation, the Resurrection, the Ascension—all are at this moment one sacramental act. I looked down into the red of the wine, mesmerized by the shimmering lights reflected on the chalice's gold bottom, at this unmoved moment of sacred space. Somehow this moment was the fullness by which to understand as foretaste the sacredness of all space. Once we recognize them, multiple sacraments and rich sacramentals surround us from all sides—even, I have come to believe, around and within a "company house."

My pilgrimage, then, has involved discovering the Church's involvement with space—for she knows how to take the materials of our common daily space and with the Spirit render them vessels of glory. Water, bread, wine, wood, ceramics, glass, paint—all marinated in the sounds of music, and enfolded within the aroma of incense. It all has to do with space. Saint Benedict in his *Rule for Monasteries* understood this expansiveness of the Church's understanding when he insisted that the tools of the monastery are to be treated as vessels of the altar.

Especially in the sacred space of the Church's worship are one's senses most transformed, and thereby prepared to participate in the sacredness of the whole cosmos. Both have to do with space—*sacred* space. William Law helpfully identifies Christian spirituality as not that of being "uncommonly good, but heroically faithful" to the vision.

To be faithful requires a special way of "seeing," in which the senses are rendered alert and intense. The formation of one's senses is imperative because, in a profound sense, sin is a matter of "taking for granted." For many persons, to see one of anything is to see them all: rain, sun, tree, moon, whatever. The deadly litany of the writer of Ecclesiastes is that "all things are wearisome." Why? Because "there is nothing new under the sun" (Eccles. 1:8, 9). He has seen it all. The modern version goes: "Been there, done that."

I work with persons on death row. In almost every case, as the time of death approaches—the days, the hours, the minutes, the seconds—something profound happens. What they have taken for granted for so long, the sheer fact of existing, becomes sacred. The final walk, the final touch, the final look, the final breath—the very last of each, within the final last of all. *Everything becomes precious.* Here we touch the nature of the conversion we deeply need. Toward those of us postponing our *present* for the sake of a *future*, God is blunt: "You fool! This very night your soul is being demanded of you" (Luke 12:20). Knowing this, that anything may be the last time, opens us to re-experience each and every thing as if it were the first. The final smile for the one you love is like falling in love all over again. Never is the ocean more blue than to one's final glance on the day the vacation ends. With this special gaze turned on each thing, every time, whatever is immediately in front of us is rendered gift-wrapped.

—⊦—

2. Looking Ahead

Time so wraps "things" that space becomes rich in varied ways. In Chapter Two, "Space-Time," we will explore how the Church, throughout its history, has employed almost every material to adorn in beauty the space of its living. "Sacramental" is the name given to any *thing* or *act* that has the power to evoke meaning that is already implicitly present. "Sacraments," on the other hand, are the acts and materials claimed by the Church as hinge points of our pilgrimage, at which the Holy Spirit promises to act in special ways. And with the sacrament of the Eucharist, we reach the primal *spatial event* of the Church's life. Here is the gesture and rhythm that unites for the Christian all the spatial dimensions with time, rehearsing and empowering the total life of the "faithful." In response, the Christian is empowered to reclaim mystery and to create spaces that are holy.

In Chapter Three, "The World of Sacrament," we will explore how reclaiming a deep sense of the Body of Christ entails understanding the interplay of "church," "sect," and "monasticism" in Christian history. It is important for us to recognize the present tendencies in Roman Catholicism and Protestantism as a reaching out for each other. In acknowledging Eucharist as the primal Christian image, we will consider ways of resolving such stumbling blocks as transubstantiation and the number of sacraments. The goal is to move from the divisive stance of either/or toward the hospitable and renewing Christian stance characterized by both/and.

In Chapter Four, "Sacraments and Sacramentals," we will begin by focusing on the two primary sacraments of both Protestantism and Catholicism: Baptism and Eucharist. By considering a number of themes emerging through the impact of Vatican II, we will identify the significant movements bringing us toward mutual understanding. With such richness, we are encouraged to recognize God's call to us as co-creators. Through our efforts at sacramental expansion, we are called to participate creatively in the sacralization and re-sacralization of space.

In Chapter Five, "The Shaping of Space as Beauty," our focus will become quite concrete. After indicating how the senses need to be transformed by formation, we will explore the primary call of the Church to transfigure space. I will share my own attempt to do this by creating a hermitage. From there we will focus on church architecture, examining seven contrasting creations of intentional sacred space. Then, turning to domestic architecture, we will explore Frank Lloyd Wright's Fallingwater, and as a model for the sacralizing of public space, we will consider the Vietnam Veterans Memorial ("The Wall"). With such a base, we will examine the spatial arenas of our own daily life that need to become expressions of sacredness: home, school, work, car, and, finally, where we die. Included will be our need for objects of beauty, pilgrimages, sacred motions, and arenas of divine visitation.

In Chapter Six, "Thoughts on God," we are brought finally to ask how our exploration of space provides new clues for understanding who God is. As Plato insisted, in the midst of mystery, we become poets telling stories. And so my story of "a descent into the Grand Canyon." Caught up in that hostile and bleak space, I came to understand the degree to which I did not belong there. In fact, I experienced to a persuasive degree that the appearance of humans appears almost as an afterthought—after eons of non–self-conscious life. The image that was birthed for me that day was that of the "reverse Trinity." If one understands God as being self-conscious from the beginning of space and time, one cannot escape the image of a sadistic God who designed and structured everything as a perennial exercise in suffering and death.

Here, in the Canyon, the problem of evil clashed acrimoniously against any Christian belief in a God of love, let alone a God who is even minimally humane. But on that warm evening, safe on the more hospitable rim, I asked another question. What if the "beginning" is God as Spirit, the One described in Genesis as moving from darkness out over the face of the deep? Such sacred restlessness, this breaking out as primal expansiveness, would then intertwine space with time as pilgrimage—as divine self-consciousness in the making. Incarnation would then be the disclosure that God struggles carnally, thrashing in and through and with the length and breadth and depth of all space, moving toward being that fully self-conscious God whom Scripture identifies as All in all. The Christ-event then is the promise within time of the transforming consummation of space. In Christ as foretaste, we experience God as the Triune Source, Companion, and Goal of all things.

With this final vision as goal, let us start at the beginning, exploring what it might mean to "Make Space Holy."

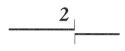

2

SPACE-TIME

"If I ascend to heaven, you are there; if I make my bed in Sheol, you are there." (Ps. 139:8)

"Your footsteps were unseen." (Ps. 77:19)

1. Clocks versus Time

There is no such thing as time if there is no such thing as space. What creates both is the presence of anything that "exists" in relationship. In a primal sense, time began when God first said, "Let there be light." In dealing with Time in the first book, we dealt implicitly with Space, for they belong together. Objects immediately create a spatial setting; and space requires senses, which for humans means hearing, touching, seeing, smelling, and tasting. The trees make wind visible. Only for noses does aroma exist. For those who are blind, there is no such thing as color. And in terms of the meaning that Christians most cherish, space-time is present when "two or three are gathered in my name" (Matt. 18:20).

Clocks *create* time of a very peculiar kind. To call clock-time "objective" is misleading. The only reality that clock-time has is for persons willing to accept such a device as useful for their "doing." If one wonders what it might be like to live in clockless time, simply participate in an African American worship service, or a Native American sacred dance, or even the silence of a Quaker gathering. What most frustrates Anglo-Americans about such events is the apparent "arbitrariness" of "beginnings" and "endings." The creation of Navajo sand paintings, the real ones rather than those on exhibition, is an intricate event within a weeklong ceremony for a person's healing. In this context, it would be sacrilegious to let clock-time intervene. An indispensable "rule" is that both leader and participants never be in a "hurry." When fatigue begins to gather, the chanting slows and quiets, until those in the hogan are asleep. The chanting begins again when "it is ready." The ceremony and the convergence of all things come together in this hogan as the center of space and time, "birthed together" when it is "right to happen."

At this time in our history especially, "mechanical time" is robbing us of time's "thickness," which is the gift of space. The only substance to clock-time is the "tick," without which such time does not exist. At retreats such as Cursillo, the participants must turn in their watches, for one is to experience space-time of another order. Mechanical time destroys the true experience of being human, of being communal, and of shepherding one's tradition. Such time is no longer the time of living what is, but of being "obedient" to a supposed reality independent of immediate experience. What windmills do for wind is akin to what clocks attempt to do for space-time, unsuccessfully. To illustrate, most tombstones meant to "honor," really don't. All that is chiseled on them along with the name are two calendar dates. The resulting image is that of time having no "thickness." All the tombstone says is that one day the person was born, and one day the person died. The appearance is that his or her "life" was one simply of enduring clock-time, and nothing of note happened.

I remember the day my family visited our favorite little "timeless" cemetery in the beautiful Crystal River Valley of Colorado. Most of the stones are old, with only faint words and dates. But over in a corner was a fresh grave. At the foot of the grave was a mayonnaise jar filled with freshly picked daisies. At the head was a two-foot plank, the end placed down in the soil. In carefully gouged letters was the name "KAREN." That's all. Nothing else— except the thickness of the name of one who was deeply loved. The calendar of mechanical time had no part. We fell in love with Karen that day, and ever since we regarded it as a special event to go and visit "her," until one visit a permanent tombstone had appeared. We refused to read even what her last name might be, or how old she was. For us she would always be "Karen." We did not return.

If one doubts the "subjective" nature of time, recall the time of youth. In one sense, it goes so very slowly. One "can't wait" to go to school, then to be permitted to drive, to graduate, to marry, to have a child, to get a promotion. "Living" is put off in anticipation of "when." Much of this is caused by carefully groomed parental expectations. The parent is probably slightly beyond middle age. Time goes faster, much faster, as one grows older—for time "yet to be" becomes less and less.

In our society, the once playful period called childhood is increasingly harnessed by competitive parental expectations—the soccer mom being chauffeur for multiple planned activities. All becomes a "doing," for the sake of a life claimed by "doing." Very early, life is reduced to "negotiated" space-time. An authentic "thickness" of, say, making music, is regarded as "wasted time" unless it is for performance or vocation. Intrinsic values, those freed from the pursuit or control of clock-time, increasingly become phased out, or are of value only as a means to something "more useful." In the spiritual direction I provide, over and over again I hear people express guilt over being involved with that which is not work or work related, plagued by the compulsion for time to be usefully "spent." Concern for the fullness of an authentic life is undercut with thoughts of selfishness for being concerned at all for one's own self. The shadow of guilt follows them heavily.

And over all of this falls sadness. The child today is largely robbed of childhood—the time of exploration, play, and intrinsic joy. Childhood should be a period lived without expectations, jammed full of the zest for living. There is never enough time to do all the imaginative things that Life proposes. And so the only limits should be the times for "coming in" and for "going to bed"— everything else flows forth from imagination. That is, until television is permitted to become a pariah of the child's imagination, and parents find their identity through their child's competitive "doing."

This atrophying of our lives by clock-time has been encouraged by the growing domination of the "scientific mind." Our attachment to quantitative research has resulted in so-called "laws of nature." Thus, just as we come to regard clock and calendar times as somehow having an independent existence, so we believe "natural law" is the determiner of space and the "objects" of space. It was inevitable, then, that over two centuries ago the "God question" became a mere footnote. What arena even remained in which a God could function, if the complex of natural laws accounted for all things in space and time? At best, an image for God might be that of Clockmaker, setting this independent complex in motion—and from that moment on God was no longer needed. In fact, there was nothing for a God to do, since "he" is unable to interfere with the clock of natural law—once started, it goes on relentlessly.

Some scientific minds today understand their task as descriptive rather than normative. Yet even description, from "galaxies" to "black holes," can give such a strangeness not only to time but also to the incredible space within and without, that the imagery characterizing our previous understandings of things divine are coming, at best, to feel "quaint."

2. Spirituality, Senses, and Imagination

The present concern for *spirituality* is disclosing the anemia of contemporary society in which our lives, lived within "thin spaces," need to take on a proper "thickness." The quantitative and chronological perspectives are useful tools in "doing," but they are not suitable determiners of the "what" and "why" of that which "is." The distinction between "doing" and "being" permits a radically different way of relating. A word increasingly used for this self-authenticating dimension of "being" is "contemplation." It refers to the disciplined ability to enter into and participate within that to which one is attending. Such a relationship has to do with the "this-ness"—for example, "red-ness"—as an undifferentiated encounter with wholeness. The intent is not to *do* anything, nor to appraise that which "is" in terms of profit, or even to scrutinize anything for "use." What is involved is experiencing "wholeness" in the immediacy of Presence.

Some years ago, when I was undergoing a harsh pilgrimage that turned out to be a search for God, I was referred to a particular nun in Denver. From the moment we met, I didn't like her. She was one of those who seemed to ooze Jesus all over the place. After a lengthy conversation, she encouraged me to attend her class. I walked into a large room with wall-to-wall carpet. At the front was a gnarled tree branch, with small lighted candles on every branch. There was no other light. There were no chairs. Instead the students were sprawled throughout the semi-dark room. After tripping over one, I sulked in a corner. I remember saying to myself: "There's no way that this nun character is going to play with my feelings!" After a period of silence, she turned on a tape. It wasn't fair. The first music was one of my favorite Bach pieces. Then the tape slowly turned to isolated sounds. The first was a rhythmic drip, then a gentle rain. The sound of water took over—first as a downpour, then the white water of a spring river, followed by waves beating on the shore. And in rhythm with the waves was the breathing of two people, in a growing intensity of lovemaking. The last sound I heard was the delightful laughter of a child.

I was undone. I don't remember if this was the end of the tape or not. It didn't matter. Against my willpower, I found myself rubbing the palms of my hands over the soft fabric of the carpet, and the candles gleamed as if being seen by watery eyes. That day I learned what I had not known before: that each of us has a primal sense. Mine, clearly, was that of hearing. In that experience, time stopped moving forward and issued into a wonderful thickness—the fullness of a timeless space—that easily laid claim to a darkened room through which the cosmos was gently flowing.

One of the significant philosophers of the eighteenth century was Anglican Bishop George Berkeley. A very simple observation launched him into a different way of perceiving things in space and time. He noted that a rock outside his study window the night before was still there in the morning. So? Our imagination has the power to create something in one's mind, which continues to exist as long as one remembers it. Using this analogy, Berkeley's question was this: How is anything and everything "held in being"? All we can say, insisted Berkeley, is that things we observe do exist, and yet they do not have about themselves the power to be self-sustaining. The solution that appeared viable to him was that all that is, *is*—because all things dwell in the imagination of God. This would make sense out of the psalmist's prayer, "[God,] *do not remember* the sins of my youth" (Ps. 25:7). The contrary prayer is "O Lord, in the fittingness of your anger, do not *forget me*."

The heart of the Christian faith is held together in this imagery of remembering ("hope") and forgetting ("forgiveness"). All things exist in the divine imagination, and being so "held" is their nature. For God to forget anyone means they no longer exist—and thus, "God, forget me not." The sins and evil of our past have no existence, not even in the divine memory. Forgiveness is identical with God's forgetting—and thus the prayer, "Lord, do not remember the sins of my youth."

The famed physicist Sir Arthur Eddington was intrigued that the further scientists go in exploring the deepest nature of that which is, the more it seems to resemble Mind. Further, we have suggested that what is most consistent for the Christian is not so much to speak of "eternity," but of "God's Time." This time relates to but is different from our time. The psalmist contrasts a thousand years of our clock-time with God's time being equivalent to "a watch in the night" (Ps. 90:4). Here it becomes intriguing to image our space as the *interiority* of God, and

God's time as the context for our time. The implication of God's imagination for our time-space, then, is to grasp how the existence of anything witnesses to the necessary existence of God. This is true not only at the "beginning" but also in every second throughout—as a whole and with each of its delicate parts. At the same time, we can begin to explore the nature of all that exists as a "something" in God's space. The task is to name the Name.

—┬—

3. Nature of Space and Place

There is mystery in everything that exists. We have just dealt with the mystery that anything exists, as well as the ongoing mystery that anything that once existed *continues* to do so. This means that one can experience both "that-ness" (that something exists) and "thisness" (what that something is) through the practice of contemplation. But the mystery goes deeper. In dealing with matter as constitutive of space, there are five ways to regard that which is.

1. *Material.* The world, and all within it, is opaque. Communication is thus by signs, arbitrarily chosen as a material shorthand (e.g., a red stop sign).
2. *Transparent.* All matter is potentially symbolic, for as created it can be seen as transparent to the Ground of its Being. Whether something is seen symbolically depends on what the culture or person

brings to the situation. All things, then, can have a symbolic quality, even down to the mystery invading every electron.

3. *Sacramental.* Although similar to a symbol, a sacramental is an act or object through which a tradition can evoke an awareness of something that is independent of the act or object itself. Sacramentals emerge largely when one intensely lives a particular religious heritage.

4. *Sacred.* This occurs through the intersection of space with time in such a way as to establish "holy space," "holy things," or "sacred objects."

5. *Sacrament.* The intersection of space with divine promise creates a unique event as a foretaste of God's kingdom.

Those who reject materialism have experienced the power of something to be more than it "is." Paul Tillich speaks of everything being potentially "transparent" to the Ground of Being, thus functioning symbolically. But because of our human condition and what we bring to the experience, some things function symbolically and some do not. Thus with symbols, we can distinguish between "then" and "now"—something can become symbolic, and then lose that quality. In this understanding, a symbol grasps us but has no content. Sacramentals grasp us because of their content.

The "sacred," in turn, has to do with the power of something to capture and hold together such anomalies as life and death, birth and burial, beginning and end, such as water, candles, crucifix, or utensils—anything that takes on an intrinsically sacred aura. The sacred then, refers to a value that is intrinsic to the object, but is so because it has been set aside by special intention. Therefore the dimension of "when" is involved.

A "sacrament" also has to do also with "where" and "when." Although Protestants tend to be unclear on this point, for the Catholic, the power of a sacrament is not due to its ability to point beyond itself (in transparency as a symbol), or its power to evoke aspects of a tradition (as in a sacramental). Instead, a *sacrament* occurs when the elements for the particular sacrament are empowered by the Holy Spirit as promised, because they are used in a particular place, by a particular celebrant, at a special time. That this will happen is based on our faith that Christ will do what he has promised. Thus bread and wine form a sacrament not because of what they are, but because a certain person uses certain words on a certain altar/table. The issue, then, is not what they *are*, but what they *become*. In contrast, both symbol and sacramental function on the premise that all of space ideally is intended to be a jeweled container for matter. Tillich gives the name "theonomy" to such a cultural situation—where almost everything is seen symbolically. Bread, in turn, can function as a sacramental rather than a sacrament by being experienced as the "staff of life," evoking our aware-ness of our spiritual "staff" being God, "from which all good things come."

In between sacramentals and sacraments falls the sacred. St. Benedict illustrates this, as we noted before, when he claimed that monks should "regard all the utensils of the monastery and its whole property as if they were the sacred vessels of the altar."[1] While this understanding implies that things are intrinsically holy whether set aside for a holy task or not, there is a type of sacredness that is more particularized. The event of the "burning bush" involves a particularity of when and where: it was likely that only Moses could have seen the bush because the purpose of the event was a particular call to Moses for a particular task: "'Moses, Moses!' And he said, 'Here I am'" (Exod. 3:4). Or another example: only Paul perceived in its entirety the vision of Christ on the Road to Damascus. In both cases, we have a special event for a special person for a special purpose.

―✝―

4. Sacramentals and Sacraments

Throughout its history, the Church has struggled between particularity and universality. The Reformation, for example, was rightly concerned with the misuse of particularized power. When sacred functioning is too narrowly defined and assigned, the danger of abuse through exclusivity arises. Martin Luther's corrective was the idea of "the priesthood of all believers." Yet, on the other hand, when sacred functioning and things become universalized, the danger is that they will be taken for granted and fall into disuse. Thus, for example, when all persons can function as priests, the meaning of priesthood can be lost. If one can confess to anyone wherever one might wish, before long a decreasing number of people will make use of that possibility. If everything is declared to be sacred, the tabernacle, for example, loses all special sacredness—and the distinction between sacred and secular becomes increasingly faint. If all locations are somehow holy, in time all things likewise become profane. Whatever it may be, if it is true of all, the temptation is that there is nothing special about anything.

In an effort to hold the poles of universality and particularity in creative tension, the Church has come to make a crucial distinction between "sacramental" and "sacrament." A *sacramental* is something that can evoke in a person or group that which is meaningful, but is independent of the sacramental itself. In that sense, "holy water" is "holy" in being set aside for a particular purpose of recall. For example, a much practiced sacramental is a gesture one does upon entering a church. One places one's fingers in a container of water, set aside for that purpose, and crosses oneself. In no way has this water objectively changed. It is no different from any other water. (Recently, however, I didn't have the nerve to tell this to an elderly woman who lives close to the church and who insists that holy water is a cure for her corns—a "liturgy" that she practices "religiously"!) Rather, the intent of this particular sacramental is to remind a person upon entering the church that this is the Body of Christ—and in Baptism we were branded, bought by Christ for a great price, and therefore are beloved by God. We can capture the meaning with six words. The wet sign of the cross is a reminder to "remember your Baptism, and be thankful!"

Sometimes a sacramental is used as if it were a sacrament—and then it is functioning as "magic." Picture in your mind a basketball game in which Notre Dame is playing Georgetown. The score is tied. As the final buzzer goes off, a Notre Dame player is fouled. As he stands at the foul line, the game depending on this shot, he suddenly gasps, holds the ball with one hand, and crosses himself with the other. Now, if he is of the mind that had he not crossed himself, God would not guide the ball through the hoop, this is no longer a sacramental. It is functioning as "magic." A sacramental has a quite different meaning. The true intent of the crossing should be to put things in perspective. It would mean for the player to recall with Paul that "if we live, we live to the Lord, and if we die, we die to the Lord; so then, whether we live or whether we die, we are the Lord's" (Rom. 14:8). Applied to the foul shooting in particular, it is a reminder that games are to be played for enjoyment. Do your best, and let it be. This is only a game.

Sacraments are different. As historically defined by the Church, a sacrament causes a change to occur. Recognizing this difference between a sacramental and a sacrament helps us identify an unacknowledged chasm between many Protestant practices and Catholic ones.

Protestant pastors often stumble over what to say to prospective parents about the Baptism of their child. "Does something happen?" they often hear. Their usual response is couched in terms of the responsibilities that the parents and godparents will pledge themselves to assume. Further conversation is likely to affirm Baptism as the congregation's acknowledgment of the sacredness of birth, and its accepting the child as a member of the Church—probationary or actual. There is nothing wrong with such a sacramental understanding. It is just that this is not a sacrament as defined by the tradition of the Church, since nothing "objective" is regarded as being changed by the act. A sacrament causes what it represents.

While funerals are no longer regarded as sacraments, the different perspectives of Protestants and Catholics can help illustrate the difference we are drawing. Many Protestant ministers prepare for a funeral by seeing this event as primarily a help for the grieving survivors. Often Elisabeth Kübler-Ross's stages of grieving are in the back of the pastor's mind. In contrast, the Catholic funeral is not focused upon the family, but on the person who died. The liturgy reaches its finale with the commending of the person unto God, assuming that if the deceased is entering the joy of the Lord, the survivors' grieving is put into perspective.

Even more pointed in contrasting diverse meanings of Baptism is a story of two Christians in Nazi Germany. One night, a pastor answered a knock on his door. An SS trooper entered and commanded the pastor to baptize his child. "I don't believe in this nonsense, but I promised my father that his grandson would be baptized." The pastor responded with a firm "No." Baptism, he explained, would have no significance unless the adults would pledge their responsibility for seeing that the child was raised as a Christian. Meanwhile, in an SS hospital on the other side of Berlin, a nurse was arrested after it was discovered that at night she would quietly enter the nursery and baptize every infant that was there. She and the pastor both died in concentration camps, for practicing their understandings of the sacrament in opposite ways.

However one might respond to this story, it is important that the Christian not minimize either sacramentals or sacraments. Tillich's approach to the sacraments rendered them, in effect, sacramentals. The "Fall" for him was the undesirable separation between baptism and a daily shower, or the shadow that falls across the functions of the altar and the act of writing a book on his desk. Likewise he found wrong the need to distinguish between Holy Communion and a fine meal with a friend. The goal of what he called "symbols" was to evoke in as wide an arena as possible the awareness that everything can be seen sacramentally, transparent to the Ground of Being—with the goal being a fully theonomous culture. Since everything is rooted for its existence in God's sustaining power, everything has the possibility of symbolically evoking that Power.

Such an approach is an important perspective for developing more fully the *sacramental* nature of authentic Christian living. But the Church, from its beginning and throughout its history, has developed an additional arena of meaning—the *sacraments*. Because of the misuse of sacraments in the period preceding the Reformation, the Reformers used Scripture as the norm for evaluating each sacrament. Although Luther, given his negative view of the human condition, was tempted to see "Confession" as a third sacrament, he came to name only two as biblically unique. A sacrament became a practice uniquely performed or taught by Jesus. The Reformer's decision that only Baptism and Eucharist are sacraments, however, is a bit shaky since Jesus requested other unique practices, such as the washing of feet.

Catholics, on the other hand, insist upon seven sacraments. The difference has to do with the norm that Catholics use for characterizing Christian life. Protestants, rooted in the Reformation, hold to Scripture alone as norm. Roman Catholics, on the other hand, regard Scripture as resulting from tradition. Therefore both Scripture and tradition are to be used, as they have intertwined throughout the Church's common life over the centuries. Catholics have a strong doctrine of the Holy Spirit, coupled with a high view of "Holy Mother Church," thereby emphasizing God's ongoing action throughout history. And although Scripture has been an increasingly important norm since Vatican II, Catholics still hold to the need to explicate Scripture through tradition—but in a manner consistent with Scripture. The result is that the Catholic Church has identified special space-time intersections at which Christ promises to be present in a special way. This does not mean that God does

not work in many other ways. It means that these intersections are those where the Christian can trust in God's
"Real Presence." Faith means trusting the divine promises.

The best way to understand the result of this interplay
between Scripture and tradition is to perceive, as we shall
see, how the seven sacraments parallel in close fashion the
stages of life. They are like hinge-points appearing at the
"crisis" forks in the road of Christian living. Here again we
encounter time, but in such a way that time intersects with
the physical objects of space, and with space itself—thereby
creating *events* as sacraments. While Protestants reject
these hinge-points as sacraments, they usually deal with
them in terms of what are generally called "ordinances."
This can be confusing, however, for the dictionary definition of an "ordinance" includes the meaning of "sacrament."
But since there is a recent Protestant tendency to treat the
two sacraments as if they were sacramentals, we can almost
see seven sacramental arenas with which Protestants tend
to be involved. The Protestant name for the Thursday of
Holy Week is "Maundy Thursday," the word "maundy"
meaning "mandate" or "command." Relatedly, the
Protestant tendency is to see ordinances as acts of obedience, while Catholics emphasize sacraments as empowering
events.

—┴—

5. Mystery

One of the most lamentable consequences of modern life is the gradual but relentless demythologizing of everything. Apparently, the assumption is that to name and to describe things "objectively" is sufficient to understand and control them. The scientific mind seems intent on removing all "mystery"—for from a scientific perspective, mystery is simply an inexact name for that which is understandable but not yet understood. I have said of myself that I have a Protestant mind and a Catholic heart. Protestantism has tended to tilt the balance toward the cerebral and the verbal. I don't know any Protestant seminary that has a class in "sacramental theology." A popular Protestant architecture is Georgian—with clear glass making indistinct the inside and the outside world. Its form is composed of 90-degree angles, circles, and triangles. Its inspiration comes from the period of Greek rationality, preceding the advent of Christianity. Everything is rational, ordered, and understandable by the mind. Protestant worship centers in a sermon that may or may not explicate the meaning of Scripture. As a result, Protestantism is dependent upon human fervor to give substance to memory.

When I was a graduate student in New Haven, I worked as a part-time reporter for the newspaper. Interestingly, the "methodology" used in both graduate school and office was nearly identical. I was trained to state things objectively, rationally arranging everything from the most inclusive to the smallest detail, which was expendable if space dictated. One evening, walking back to campus, I passed a Catholic Church that I had hardly noticed before. There was a side door, at street level, that was slightly ajar. As a Protestant I was brought up to have disdain for Catholics, and little more than sympathy for their bizarre beliefs and practices. Could I sneak in the side door and just look, without being caught? I had never been in a Catholic Church before. So I did. There in the growing shadows of an autumn evening I entered a world of mystery—candles of every color, figures that could or could not have been "real," and pleasant aromas. It was an arena of silence, until this world creaked as I eased into a pew. Marvelous sounds—off in a distance, I heard the muffled traffic of night. Windows sparkled as if jewels. For at least a moment, I lost the need, or even the care, to know or understand. Here I could just participate in Mystery. It was a "World of Sacrament." The steps back into the "other world" were strange—walking the streets toward what I had previously regarded as "home." Little did I know how and by what I had been bitten.

I celebrate the epoch-making event of Vatican II. Without it, I likely would not be on the pilgrimage that I am. Very important were its efforts to give clarity of meaning to the whole and to its part. And yet I have some anxiety about the loss that excessive rationality can have on both liturgy and the sacred. The tabernacle once was reserved for only priests to open; and the sanctuary holding the sacred objects was a place reserved for priests. The wafer that was held up before one's eyes with the words "the body of Christ" was so sacred that the only touch was as the priest put it in one's mouth. Even a dish was held underneath one's jaw, just in case the host might drop. Since Vatican II, however, there are now lay "Eucharistic ministers" who are free to open the tabernacle. They give communion, and are able to carry the host in a small pyx in one's pocket, taking it to the sick. Lay persons pour wine at the altar from cruets into chalices, as the priest leaves the sanctuary section of the church for the passing of the peace within the congregation. Don't get me wrong. I affirm these changes. And yet, as a cradle-Protestant, I know the alligators that lurk down that path.

What did I feel in climbing the huge flight of steps in the Yucatan, climbing carefully above the jungle below and entering a silent temple on top? Where are the places any longer where one can be haunted by an emotion hardly felt since childhood? What did I experience in climbing a tree to my hideaway—in order to cry? It is mystery, the mystery of the deep caves in southern France, where by torchlight our ancestors sketched on the womb-like walls the mystery that united animate and inanimate, sky and earth. Joseph Martos understands that the gain in understanding has been at the price of a loss of Mystery. The Latin Mass before Vatican II was like a secret. The priest whispered in a "mysterious language" over "hidden objects on a high altar." The music was from a different age and seemingly from a different world. The incense and pageantry were for the more festive celebrations. The feel was that "something supernatural was occurring." Likewise confessions were a haunting mystery, where in a black "box" an unseen voice spoke, almost as if it was God. This voice asked us to trust, to disclose everything that one tried to keep secret. And at the baptistery, "one had the sense of a soul's being rescued from eternal sorrow." Meanwhile, weddings and ordinations were so mysterious that they established a change that was indissoluble for life.[2]

Having said all this, liturgical renewal has made great strides in opening laity to share in the mystery. But where, now, can laity, ministers, and priests alike, be drenched in mystery? I would venture that the present fascination with labyrinths is an attempt to address this longing. Since daily life is composed of mazes, a labyrinth refreshes the soul when one traverses a long distance physically in order to reach a center that all the time was quite close, on a path where one could not get lost. What are we to make of a society that, having "everything," experiments voraciously with drugs of all kinds? Apparently we are in the midst of intense efforts to subjugate the negative feelings that haunt us from the mysterious unconscious. Or is our desire to enter a state that might be within hailing distance of what the mystic Gregory of Nyssa called "sober inebriation." A well-known monastic contemplative once told me, "In a society that emasculates mystery as ours does, I grow my own marijuana." Given the entertainment orientation of what our present-day pundits define as "news," there is hardly anyone who any longer has an aura of mystery about them. No one is left whose shoes are without mud. Even the mystery of soul in our "saints" is being psychoanalyzed as a messy sublimation of sex and otherworldly greed.

While our recent fascination with "spirituality" may indeed be, as is often said, a desire to "seek God," it is probably more accurate to identify it as a vigorous effort to break out of the so-called parameters of the "real." In my work as spiritual director, I encounter in others a hunger that clamors for "mystery" as an indispensable dimension of human life. Actually our dilemma may be deeper than our apparent disregard for things once "holy." We are apparently incapable of perceiving mystery in the sheer fact that anything exists. Society has deprived us of a spirit capable of seeing mystery in the obvious. The world of matter is utterly mysterious, for the very nature of what we perceive with our senses is not at all what a scientific perspective calls "real." What appears solid is, scientists tell us, just a porous frenzy of electrons, and what appears to function by "law" actually functions, at its deepest level, as a chaos of the indeterminate. What is solid, firm, still, and colored, turns out to be, as the scientists claims, the very opposite—transparent and invisible to the core. When one perceives how amazingly different are these contrasting ways in which space and time can be perceived and imaged, one realizes how presumptuous it is to settle on one as the determiner of the "true."

How tragic that our education is heavily focused on the "objective" studies, from physics to chemistry to mathematics. Even these are focused upon use in production. When educational funds are short, as often seems to be the case, it is the "fine arts" that are canceled. With that supposedly virtuous setting of priorities, the sensibility of the young is forged in such a way that "exile" results. What of the poetic soul in each of us, or those who dance gracefully with the angels? The result is a modern "mind" that is incapable of understanding a Cezanne, who spent much of his life painting the same mountain—for reasons deeper than a regard for its mineral content. I shudder to think of young life formed so as never to be lured by great music, or wrapped in the secret silence of wilderness, or startled by the smell of honeysuckle at dawn, or at one with a kayak aimed toward the sun's rippled setting. In Mozart's day Bach was largely dismissed, to Mozart's amazement. I would like to believe that the rediscovery of Bach in our time is the appeal of baroque creativity, which promises a Light drawn forth from a delight in tensions and contradictions. Handicapped is the person who never discovers that there is a depth of Spirit that only great music can touch. And I would be considered quite weird indeed if I declared poetry to be the index for college entrance.

Yet the technology of excellent sound reproduction, and the amazing availability of art in all forms through the Internet, give hope that such experience may even yet be more available. In fact, as I write, my computer is playing for me Handel's "Oboe Concertoes." And meanwhile, scientists are sounding increasingly like poets as they describe their results. There is hope.

Paradox runs the whole gamut of life, crying out to be grasped more than resolved—the weak/strong, low/high, sterile/rich, actual/possible, within/without, hungry/full. The Church attempts to be enfolded by such Mystery in what is called liturgy, which is the spiritual intertwining of space and time. Liturgy operates mostly in terms of the whole, the large, the inclusive; it occurs at the high points within the dynamics of human life. But this large brush for painting the anatomy of space and time is increasingly difficult for mobile and self-occupied Americans to grasp, even though some are finding it as one of the few constants in life. Thus there is something to be said for Annie Dillard's proposal: to recover mystery in the small everyday occurrences. With this understanding, patting a puppy takes on the mystery of infinite importance. I too am convinced that taking a drink of water, if performed knowingly, is to be in touch with the divine. Gerard Manley Hopkins draws forth this meaning in words: "The World is charged with the grandeur of God … like shining from shook foil." [3]

—

6. Blessing: Naming the Name

Just this past week I was priest for a sacred event that had been scheduled a year in advance. The couple was rather indifferent to the liturgy of the local church, but this event was that of blessing their new home. It was so important for them that relatives and friends from as far away as both coasts were present. The small children, expecting to be bored, grouped themselves together in the far corner of the living room, poking at each other. The liturgy I had planned, however, was mobile. We began with blessing the home from the outside, moving from there, across the threshold, to bless each room and part and function of the house. I had a bowl of "holy" water for this blessing, and invited the children to help me with the sprinkling. They competed loudly to be the one to bless the next room. In fact, they kept finding more and more things to bless—including tricycles, a dollhouse, and two dogs. And for the incredible brunch that followed, it was the children who suggested a blessing, sprinkling each plate, this time with milk.

A friend told me recently of attending Trinity Church in Boston when Thomas Shaw, the Episcopal Bishop of Boston, was officiating at Baptisms. It turned out that his custom was to have all the children come forward and gather around the baptistery. Then, during the blessing of the water, he had all the children help him stir it.

In 1987, the Catholics published an English edition of a large, official volume entitled "Book of Blessings."[4] It is divided into six parts: blessings for *Persons* (e.g., families, children, childbirth, adoption, sickness, meetings, birthdays, travelers, miscarriage); *Buildings and Human Activities* (e.g., sites, homes, seminary, university, library, hospital, means of transportation, computers, tools, animals, seeds, harvest, athletic events, meals, and even boats and fishing gear); *Objects* (e.g., baptistery, oils, tabernacle, confessional, doors, cross, images, wells, organ, chalice, stations, cemetery); *Articles* (e.g., rosaries, scapulars); *Feasts and Seasons* (e.g., Advent wreath, manger, Christmas tree, homes, ashes, table, Easter, Mother's Day, Father's Day, All Souls, Thanksgiving, and even devotional food); *Various Occasions* (e.g., persons involved in pastoral service, readers, servers, musicians, ushers, eucharistic ministers, parish council, new parishioners, departing parishioners, ecclesiastical honors, installing a pastor, blessing of the whole people of God). Other than mentioning by name goldfish bowls and doghouses, just about every aspect of a Christian's life is to be consecrated. And this entire liturgy is outside the regular liturgy of the Church.

The syntax of a blessing is this: "May the Spirit bless…." What makes these requests appropriate for the Christian is that the whole creation was blessed in that holy intersection of space and time entitled "Jesus." This is why the prayer of Christians ends "in the name of Jesus Christ," for everything depends on faithfulness to the God whose promises bless us.

Edward Hays's comparable book is wisely called *Prayers for the Domestic Church*.[5] Luther, by emphasizing "the priesthood of all believers," moved the center of his spirituality from the life of monks in community to the domestic church known as the Christian family, which has been central for Protestants ever since. Even in my boyhood, our family had a special "devotional corner" in the living room where we daily read the Bible together. It was where we placed mementos of our life together. And it was in the large family Bible that we wrote the important events of each person's life.

Both Catholic and Protestant traditions presently are experiencing an expansive rebirth of Christian life as more fully sacramental. Both are intent on viewing the world as jeweled and radiant, colorfully alive for those whose sensitivities are honed to see, and hear, and touch, and smell, and taste. All that is, in every arena of doing and being, from home, to job, to place of worship, to leisure—all people and all things and all places are to be blessed—for their very being is a gift, their use blessed, and their meaning sacred. So the Church has been through the ages—in the architecture that encloses our seeing, the music that our hearing carries to our souls, the touch of vestments and water, the smell of fragrant incense, and the taste of domestic wine and freshly baked bread. This is, in truth, the effort of tradition through which the Christian may experience in depth the foretaste of the promised Kingdom of God to come—now.

Whether guided or misguided, an argument persists in our country over whether to make the "desecration" of the American flag punishable by law. Such persistence illustrates the power and claim that symbols can have in our lives. During the 1960s a professor friend of mine confessed his embarrassment one day as we walked to lunch. It appeared that when one filled one's car with gasoline at a particular station, they put a free American flag on the radio antenna. Just then, we reached his car, adorned as if on its way to the Republican National Convention. "I can take care of that," I said, as I pulled a pack of matches from my pocket. As if from some deep wellspring, my friend struck my arm so as to send the matches across the side-walk. "Don't you dare!" he shouted. Thoroughly confused by his own behavior, he went back to his office, having lost his appetite. Expressed more theologically, one can experience each and every thing, all portions of life, as being transparent to the Ground of Creation, splendidly adorning the world as a Christmas tree. And in spite of what one might say about consciousness, the power of symbol infects deeply the unconscious, with surprising power.

I already mentioned the nun who oozed Jesus. Not only did she trap me into having my eyes redeemed by sound, but she did something else equally important. In looking at my life, which I regarded as anything but sacred, she asked me, "What do you do for Vespers?" That had an easy answer. "Nothing." "Well, what do you do when you come home from work?" "Well, I put on some music, check in with the family, and then we have a glass of wine together." She laughed. She shouldn't have. "Do you realize that this is almost exactly what we nuns do at Vespers in our monastery! Who on earth robbed you of the ability to name the Name?" To get back at her, I made a mean comment about their patron saint. "Saint Teresa had orgies with Jesus every day. Is it too much for me to experience God just once in a lifetime?" Her response was prompt, and well aimed. "Have you ever had any moments of ecstasy at all?" "Of course," I answered, a bit incensed by the question.

So I shared my feelings when I listen to Mozart. She laughed again—and I had not yet forgiven her for the first laugh. "The way you describe the experience of music is precisely the way Teresa describes her 'orgies' with Jesus!" Then came that question again: "Who robbed you of the ability to name the Name?"

I did not know who or what had robbed me. Part of it might have been in being raised Protestant, where the use of matter to express spirit was put down as "Catholic." And yet as Protestants we had our own "sacramentals." Our prayer book was the bulletin, our "idols" were the trophies of the church's sports teams, the "incense" was mildew in the basement Sunday school rooms, the sounds were off-key choir anthems and the husky singing of hymns. And for unequaled fellowship, nothing matched our potlucks. But after Vatican II, as Catholics began moving their statues to the back of the church, Protestants began using ashes and vestments to supplement their verbosity. And while Protestants continue to have problems with such doctrines as "transubstantiation," the work of such Catholic theologians as Teilhard de Chardin is helping us to see how the "scientific" view of evolution can be opened to Spirit. Such "sacramental theology" can discern how "transubstantiation" has been going on since the beginning of time, incarnating matter with Spirit.

Catholics and Protestants may be approaching hailing distance. Liturgy can now be recognized as the distillation of this expansiveness of time and space, imaging *the event of Jesus Christ as the Symbol of symbols, the Sacrament of sacraments, the end within time, the human as divine,* with space and time being the arena of God's most cherished work and play. Even the eyes of the scientist might be lured into perceiving matter as condensed energy that dances. Or what can be more mysterious than the workings of astrophysics, where only arbitrary mathematical signs are possible in the face of gazing into 100,000 billion galaxies, with 100,000 million stars in a galaxy, and the likelihood that there are an infinite number of other galaxies, whatever that might mean. Perhaps such immensity will be enough to entice theology to change. While the task of theology was once that of accounting for the solid givens of the observable world, we have reached the point of history in which *theology, to be significant, must be written by those whose senses are poetically charged, and who have a firm sense that music is preferential to all but silence.*

—

7. Holiness

We have considered the view that material is opaque, useful in an arbitrary naming of things. But if this becomes an exclusive perspective, the result is a parochial "materialism." The second perspective named was that of "transparency." Here material things may be experienced as symbols, giving a transparent feel to the material sphere. We developed as well the distinction between sacramentals and sacraments. We need, now, to recover the mystery potentially resident within the Church's interior tradition. The word *holy* refers to "religious" places and things. As we have seen, everything in creation can be viewed sacramentally. But both the "holy" and "sacraments" have an "objective" quality about them. Unlike symbols, they are not dependent upon the subjectivity of the perceiver, but on the promise of God, as authenticated by the Church. To move from symbol to the holy is to conceive matter within space not in terms of "transparency," but as "sacred."

Societies have always felt a need to commemorate the "sacred deaths" that have occurred in the past, commemorating the society's courage and valor. Memorials on most town squares, at least in the Midwest, are of the "sign" type. That is, on a piece of stone appear a few words of tribute and identification, dates of the particular conflict, and perhaps a statue of generic likeness. But two national monuments have the hallmarks of the sacramental, perhaps with the power to create "holy ground": the Lincoln Memorial and the Vietnam Veterans Memorial ("The Wall"). In both cases, interestingly, it is in honoring the particularity of persons, by figure or by name, that there arises a claim to an almost universal sense of the sacred that comes close to holiness.

"Holy" can be applied to a number of arenas. It can apply to a day, place, time, event, thing, or person. It differs from the symbolic in the sense that it is authorized by a religious tradition or ecclesiastic official as *being made holy*. We can illustrate by analogy how this usually comes about by looking at the Roman Catholic doctrine of papal infallibility. This doctrine does not hold that anything a pope might say, even about "faith and morals," is infallible. It applies primarily to that which is pronounced not *to* the Church but *as* the Church. Highly questionable, then, would be any papal pronouncement that would raise from a significant portion of the "people of God" the exclamation, "Oh, no!" or even, "Where on earth did that come from?" While infallible pronouncements are extremely rare, some papal messages, such as the supposed impossibility of ordaining women, attempt nevertheless to hint of infallibility by declaring that the argument is at an end. Instead, in this case as an example, the argument becomes even more widespread.

Efforts by the hierarchy to change the mind of a significant portion of the Church not only will not work, but seriously undermine the credibility of the papacy. On the other hand, an utterance that can lay claim to "infallibility" would be one to which a significant portion of the grass roots would respond, "Precisely; wonder what took the papacy so long to pronounce it!" Holiness, liturgy, and sacrament are phenomena that have arisen over the centuries in a manner apparently spontaneous, but understood to be the workings of the Spirit. In time, such matters are officially approved, done so when it is clear that the papacy is responding *for* and *with* the Church, and not *to* it or *at* it. In fact, there is good reason to understand the decisions of Vatican II as happening in this same way.

In the Hebraic understanding, the "Holy" focuses on the Temple and tabernacle in general, and the Holy of

Holies, where the Ark resided, in particular. Yet through-out the Old Testament, we also encounter holy things, holy ground, holy places, holy persons, holy acts, holy spaces, and holiness as an attribute. The "requirement" for being so regarded is that they can be set aside and consecrated in such a manner that their common usage is surrendered for a use sacred in meaning. In the fullest sense, God is the Holy of Holies, set aside by his infinite perfections, whose acts are thereby so holy that God's holiness can fill the earth (Ps. 24:1-2; Isa. 6:3). Even so, to be affirmed, holiness must be in a concrete intersection of space and time, where, for a time of God's choosing, God's Holiness is pleased to dwell as Presence. Thus in Catholicism, for a place such as Lourdes to be regarded as holy, one affirms a "special visitation" as certified by the miracles that have occurred. Likewise, a deceased person can be declared "holy," i.e., a saint, after careful determination that several "miraculous" responses have occurred through prayers to that person for divine help.

Once, while walking through the British Museum, standing finally before the Elgin Marbles, I became almost overcome by the travesty of that world-famous museum. My perception centered in an awareness of the vast difference between secular and spiritual *memory*, in the contrast between being a tourist attraction and being a place of holy pilgrimage. People once regarded much of what is now in that museum as holy. Whether one is viewing a frieze from a Greek temple or the mummy that was once an Egyptian god, worship is utterly absent, and the past as "holy" has been negated, for that which is of "historic" interest.

Evangelical Protestants tend to have no acknowledged saints, no holy places, no sculpted figures, no tabernacle, no liturgy as drama. As a result, "Presence" often seems absent. Instead, the penchant is to preach and teach *about* that which is divine. It may be indicative that silence is mostly eliminated, and even the word *presence* is usually an unfamiliar one. Mainline Protestantism tends to approach Scripture historically, as "Bible study." This is in contrast with the monastic approach, called *lectio divina*— a method through which one becomes so immersed in a biblical passage that God speaks immediately and intimately through it. The piety of their respective traditions is not generally appreciated. In attempting to attain a compatibility with science, a number of scholars, mostly Protestant, have continued the "demythologizing" program of Rudolf Bultmann. An alternative tradition is one that focuses on Presence through participating in it. Surrounding the poetry of liturgy are evocative tapestries, carvings, paintings—the craftsmanship of dedicated and accomplished artists. For rediscovering this tradition, Protestantism can well look within itself at the leadings of the Anglican Church.

On a more practical plane, intercessory prayer tends to fade from both worship and the personal spirituality of liberal Protestant pastors, for the idea of *changing* God's mind is theologically unacceptable. Prayer, then, as Tillich contends, is not really intended for God, but for the one praying—that one's mind might be made to accept God's will in all things. The change is in us.

In contrast with some of these tendencies of modern Protestantism are the Protestant Reformers themselves. Luther and John Calvin spent hours in daily prayer, firmly believing that a significant Christian practice is to *contend* with God—in anger, pleading, urging, threatening—as in Jesus' parable of the persistent widow who hounded the judge so frequently that she finally got her way (Luke 18:1-8). So it was with Jesus himself, who for hours pleaded with God to remove the "cup" before him. Three times he returned to his solitary prayer, contending with God, with prayers so intense that his sweat was like drops of blood. In heavy contrast to this portrait is a God functionally bound by "natural law"—one that is neither scriptural nor even interesting. At the heart of Scripture is a God who can and will do as God promises, and Jesus Christ is the one who discloses these promises.

The Old Testament is filled with time-space intersections that, because of what happened there for a person, were declared to be holy. This is so from the place of Jacob's dream of the heavenly ladder, through the numerous wells dug by Israel's important figures, to Solomon's Temple. A difficult issue in coming to terms with "holiness," however, is the issue of *permanence*. Is it the case that what is declared "holy" is intrinsically so, or is it an ascribed "holiness" that remains for a time?

—⊦—

8. The Pilgrimage

One clue for considering the permanence of "holiness" is found in the image of "pilgrimage," which was once an important spiritual discipline and to some degree still is. Catholicism has a number of locations that in being called "holy" are places of pilgrimage. The major pilgrimage is to St. Peter's in Rome, the citadel of "Holy Mother Church." Protestants, on the other hand, who often substitute exhausting "vacations" for "pilgrimages," recently have been coming to seek a pilgrimage after all—a visit to the "Holy Land."

In calling these two places holy, it is important that they not simply have historic or even religious interest, but are holy in the sense of being recognized as places of divine encounter. To understand this, we need to struggle with what the Church has meant through the ages in encouraging pilgrimages to places linked to the history of salvation. The Christian will no doubt acknowledge the power of the Spirit to blow where it wills, using as instruments places and people of God's own choosing. Yet there is little question but that the God of the Scriptures chose various places to "pitch his tent" (e.g., Exod. 40:34-35; 1 Kings 8:10-13; John 1:14). But what about "permanent" locations of holiness?

Pope John Paul II was helpful recently in reclaiming this tradition. He declared that the spatial dimension is no less decisive than the temporal in the concrete accomplishment of the mystery of Incarnation. Christians consider the whole world to be the citadel of God's glory, indeed the temple of God's Presence—and yet there is more. In the first book, we spoke of *kairos* in regard to time—those special moments when God is particularly present for God's own particular reasons. Therefore we can meaningfully speak of Jesus as the *Kairos* of all *kairoi*, that special Incarnation that appeared in the "fullness of time." This was not simply once in history, but Jesus is the *Kairos* for all of time. Thus just as there are special times of grace, by analogy we can acknowledge space as bearing the stamp of God's particular saving actions. In one sense, Jesus could have been born anywhere, and yet to be who he was, and for what he was to do, Israel was the place. Not surprisingly, Scripture goes to great lengths to show that by ancestry, birth place, and place of ministry, this was indeed the one who had been prophesied.

Many places came to be recognized as holy because of Jesus' presence there. Holiness is more than sacramental. It is more than symbolic, in transparency to the Ground of Being. Holiness is the "incarnation" of Spirit with matter, so much so that it is a quality taken on by that "matter," not intrinsically, but for as long as the Spirit wills it. Such holiness is not generally known until it occurs, as the Spirit goes to and fro as it wills. Yet a holy place or object has such intensity that one might well assume longevity for that Presence. It requires the syntax of "here" but not "there."

For those skeptical of the Spirit's "intrusion," I find no better evidence than the major Ecumenical Councils. They were thoroughly political events, exercises of power,

with people jockeying for position. Whatever their decisions, on those decisions Christianity would rest. And yet
from them came the incredible decisions on which
Christianity does rest, as in the case of the Nicene Creed
and the decisions at Chalcedon. Clearly a "nevertheless"
was involved.

Places regarded as "holy" are locations where
encounter with the divine may be experienced more
intensely than normally experienced in the vastness of the
cosmos. One can gain a sense for the "holy" by visualizing
what our spiritual life would be like if expressed in terms of
places and actions. For example, I am not the kind of person who could be called emotional, trusting more on my
reason. Several years ago, after attending an international
conference in Africa, I decided to make a side trip to
Jerusalem. Friends had cautioned me about going, saying
that my rationality would be offended by the appeal to
tourists to visit arbitrary sites declared "holy" for purposes of
making money. I went anyhow. I arrived by bus on a Friday
around 4:00 PM. Having no idea what to do, I saw a sign for
the "Catholic Information Office." That was worth a try.
Right inside the door was a poster for the Via Delorosa, the
"Way of the Cross," led by the Franciscan Brothers every
Friday at 3:00. I'd missed it! Sadly, I asked the pleasant lady
what I should see. "The Via Delorosa, for sure." "That was an
hour ago." "No, our time is earlier than Tel Aviv."

Armed with quick instructions as to how to intersect
the procession at the second station, I ran very impolitely
through back walkways and steps. There they were—about

seventy-five "pilgrims," walking the supposed path of Jesus on the fourteen stations to his death. As we moved through the narrow walkways lined with merchants of every kind, there was clear animosity between Jew and Christian. The old Jerusalem had about it an aura of never having changed since the time of Jesus himself—both in place and in dress. Feeling gentle about the harsh words and shoving we received from merchants, we finally came to a door of a particularly old church, kept standing by I-beams and wire. Up a pair of steps we went, after a sharp right. At the top were lighted candles and lamps of every type, size, and color. After the prayer for station twelve, people began descending by another staircase. But before leaving, each person knelt in front of an elaborate altar and put their hand in a hole in the rock floor. "Where are we?" I politely asked the man beside me. "Golgotha. That is the hole where Jesus' cross was placed." Bewildered, I took my turn, and with my hand in that hole, this skeptical academician felt what I had never quite felt before. Far more was going on in me than tourist curiosity. I too went down the stairs, far more thoughtfully than when I had gone up.

By the time I caught up with the group, there was a line in front of a small cavelike structure in the center of this large church. People went in and out, one at a time. My time came. I stooped low, and entered. There was a small room, with a cavelike hole into the next "room." In it, there was what appeared to be a bedlike bench carved from the stone. In the candlelight I saw a Franciscan monk seated in the corner. The two of us almost filled the space. "Where am I?" I timidly asked. "This is Christ's tomb—the tomb of the Resurrection." I couldn't move. Finally I did, leaving on all fours. I went around behind the edifice, where in the darkness I could cry. I felt full to overflowing. All I could say was "This is a holy place." A monk saw me in the darkness. He came over, and pressed into my hand a rosary carved of wood from the Mount of Olives. "You need this." That was all he said. As I went out into the twilight of the "Eternal City," I clutched the rosary. That was when I learned what a holy place is.

Protestants are inclined to be skeptical about such "holy" places, especially when they have anything to do with the so-called visitations of Mary. But Protestants themselves show the universal attraction to places that are more than places of nostalgia or sentimentality. Many people experience a heavy pull to visit the house of one's birth, and to see if the apple tree is still in the back yard—the one they climbed in order to dream. It may be almost too much to stand at the spot where one first met one's spouse, or to walk the rickety steps up to one's first cold-water flat, the day still feeling like one's honeymoon. Or there is the hospital delivery room where one experienced the miracle of the first child. Or one can stand in the cemetery where the remains of one's parents are buried. All of us have favorite records, and books, and a chair, and pictures, or even that special screwdriver that can open anything. The so-called clutter in the basement, attic, and garage does not necessarily signal laziness. And while the clutter does reflect our society's materialism and obsession with "stuff," it also discloses something deep within each of us. It is the harbinger of sacredness, of what is almost holy for us about certain things from the past, a holiness intensely present in a particular picture of a person, a place, or a thing. Our obsession with cameras is an obsession with stopping time and hallowing space. And somehow "holy" is a word we might dare to begin using again. If indeed our God is the Lord of resurrection, the "holy" indicates our nominations for eternal life.

Nevertheless, Scripture warns us about the risk of holy space. It can become idolatrous, especially if we divinize nature in part or in whole. Somewhat in response, sacred spaces in the Old Testament gradually came to be "concentrated," primarily in the Jerusalem Temple. "I was glad when they said to me, 'Let us go to the house of the LORD!'" (Ps. 122:1). But soon after the death of Jesus, the Temple was destroyed. How natural it was, then, for the first generation of Christians to see in Jesus himself that intensity marking the very residence of holiness. "Destroy this temple, and in three days I will raise it up" (John 2:19). The step was a small one for Paul to see every Christian as a temple in which the Holy Spirit through Baptism takes residence. "Do you not know that you are God's temple and that God's spirit dwells in you?" (1 Cor. 3:16). If each of us is "holy space," living the "time" of our lives, how appropriate to understand our words and actions as intended for holiness. Christ has reconciled us "so as to present you holy and blameless and irreproachable before him" (Col. 1:22).

9. The Divine Promises

What the early Church gradually realized was that Christ's words and life were promises—of two kinds. Christ promised to send the Comforter, the apparently unpredictable One, who sent some people into the desert, while others acted as if they were intoxicated at breakfast time. One can trust the Spirit, this Spirit-without-a-program, for this is the One sent by Jesus. So much are Jesus and the Spirit related that Paul could call "it" the "Spirit of Christ."

Christ's second promise was the founding of the Church as his Body, a Body against which the powers of death will not prevail, and which will have the keys of the Kingdom (Matt. 16:18-20). In this passage, the power of forgiveness is named. But throughout the Gospels, other actions are identified for the Church to do, under the promise that the Spirit himself will be the empowering One. *These promised intersections are what the Church calls sacraments.* Jesus "is the one who baptizes with the Holy Spirit" (John 1:33).

The Church is the "*World of Sacrament*" resting on the promise that the Church is present "where two or three are gathered in my name" (Matt. 18:20). Although God is present everywhere and not restricted to any one place, still the God of "no place in particular" is the one for whom "presence" and "absence" are indistinguishable. The Christian God is One who comes and goes, yes. But this is also the God who takes sides, the One who is the maker of covenants and promises. And just as Jesus Christ was the divine-human event in which God was supremely Present, so it is in turn that the Church is the Body of Christ in which Presence is supremely Present. In praising Abraham and Sarah for their faith, Paul said: "No distrust made him waver concerning the promise of God, …fully convinced that God was able to do what he had promised" (Rom.4:20-21).

"Symbols" are objects or events of transparency that depend upon the subject to see what was always there to be perceived. A "sacramental" is an object or gesture that evokes in the participant an awareness of something that is already true, but in need of something to evoke remembrance. A "blessing" is a prayer request for the Presence of the Holy Spirit with someone, at some place, or for some reason. Sometimes that Presence is so intense that the word "holy" may be authorized—and while it may be perceived as having come and gone, the Presence may also be perceived as continuing, deserving visitation by the faithful, trusting the Spirit promised by Christ. And finally, "sacrament" is the term given to an action done by the Church in which Christ has promised that the Spirit will be present, to will and to act.

The story is told of the old rabbi who once during a crisis went during a crisis to a sacred place in the forest, lit a sacred fire, and said sacred words—and the Jews were saved. The next rabbi at the next crisis knew the place and the fire, but not the words. It was enough. The next knew neither the fire nor the words, but remembered the place.

This too was enough. The final rabbi confessed that he knew not the words nor the fire nor the location, but he remembered the story. And it was enough.

In spite of the power of that story, I do not think that for Christians, telling the story is sufficient. We need not only the ability to tell the "old, old story," but also the ability so to participate in that story that the Presence of Christ is experienced as an ongoing reality. Put another way, this reality is known not only by the Word proclaimed, but also, and necessarily, by the sacraments administered. The heart of holiness for the Christian, on which any other intersections of space and time depend for their "holiness," are those acts through which *anamnesis* is possible (re-remembering as presence), and with which *epiclesis* (invoking an action by the Holy Spirit, resting on the divine promise that it will be so) is appropriate. This is the foundation for understanding the Church as the "World of Sacrament." Without divine Promise, that which is "holy" is little more than a place or thing that was *once* regarded as "holy" by someone else, in whom we have an historic interest—or none at all.

—⊢—

10. The Hermit on the Hill

My exploration of the meaning of "holiness" or "the holy" centers on a monk named Fr. Robert. He is my spiritual director. He lives in a primitive one-room cabin on the top of an Ozark hill, without running water or electricity. He rises at 1:30 AM to pray, becoming lost in the Presence of God. This time of rich participation in the divine Presence is "interrupted" only for two simple meals and for the same daily offices that all Trappist monks observe. Through him I have gained my best sense of holiness—as the residence of the Spirit in one who needs very little more. He is a gently happy person, much like a boy at play. At daybreak, he celebrates the Eucharist at a small desk in front of his window, from which he can see the hills rolling, one after another, from green to hazy blue. This is the time when he lifts up in praise and thanksgiving every living thing into God, that all may experience the Spirit which is Holy. He knows what he is about. "All I want is to grow old loving my God." The monastery tells the story that not long ago a Trappistine convent wrote to Fr. Robert asking him to become their chaplain. Robert handed the request to his Abbot, to seek discernment from the monastic community. The response was a unanimous no. "We need our hermit on the hill lest we forget why we are making and selling fruitcakes in the valley." This reminds me of a picture on my desk of a tiny Ethiopian shack on top of a steep green hill. I am told that on many of the hills there, a hermit is dwelling.

There is an old Jewish belief that there are a handful of holy persons, unknown to all but God, because of which God does not destroy the world. It rests on Abraham's contention with God, leading to a wager that God would not destroy Sodom if ten righteous persons were found

(Gen. 18:16-33). I call this the "sacrament of the few." And I think "the few" are not only those willing to die for the faith, but also those in the hills who have given up all except prayer—on behalf of us all.

—⊢—

11. Holiness and the Wilderness

The contemporary interest in ecology is becoming more than a practical program to conserve nonrenewable resources, important as this is. The movement is increasingly focusing on the preservation and expansion of land whose nature is such as to be set aside as "wilderness." The underlying assumption seems to be that when all lands of exquisite beauty are domesticated with RVs sufficient to resemble urban sprawl, something profoundly and incredibly important to being human will have been lost forever. The wilderness dare not become an "endangered species," for it does for us what nothing else can do. We can call it "secular holiness."

John Muir has done more than any other person to preserve the wilderness in this country. At the heart of his practical efforts was the spiritual quality of a gentle vision. From somewhere, I copied down his words: "Everybody needs beauty as well as bread, places to play in and pray in, where nature may heal and give strength to body and soul alike." He understood the holy intersection of space and time. One can almost hear the cries as oil-spills invade the sacredness of the ocean, and ski lifts bruise the forests we had thought to be protected "against humans." Even those who are skeptical of "holy places" such as Lourdes, or Guadeloupe, or Fatima, or pilgrimages to the graves of saints—even these persons cannot deny how particular places can remind, replenish, and heal us. As one who loves the feel of climbing the 14,000-foot peaks of Colorado, I find incomprehensible those whose perspective has no place for this mystery which heals. Spirit is at the depth of matter, mysteriously present as if a "soul."

Isaiah's classic spiritual experience warns us against "domesticating" Holiness. The context for Isaiah's experience was a moribund society, worshiping expensive finery, bracelets, headdresses, perfume boxes, festal robes, handbags, and nose rings. (Isa. 3:18-25). It was against such a culture that God appeared. As the seraphims shouted "Holy, Holy, Holy," the whole earth was filled with God's glory. The foundations of the thresholds shook, his voice thundered with smoke, for nothing less than burning coals could change the prophet's uncleanness (Isa. 6:1-9). In alarming contrast is the "God" of the modern church—polite, to a fault. Dorothee Soelle senses the loss: "God, but I want madness! I want to tremble, to be shaken, to yield to pulsation, to surrender to the rhythm of music and the sea, to the seasons of ebb and flow, to the tidal surge of love."[6] Craving to be reborn is the wild holiness of space and place—of mountains, waters, deserts, hills, and plains. It is indicative how many times the words "holy" and "ground" appear in Scripture.

The "holiness" that needs to invade the amiable spirituality of today is something of the untamed passion of the 1960s. During that period, a task force investigating the spirituality of seminaries proposed something we should not dismiss lightly. In essence it said that the more spiritual persons are, the more carnally they need to live. Denying the dualism of spiritual and material, the task force proclaimed that a student needs to see what "the wine he drinks at Communion has to do with the beer he drinks downstreet." One must see that the "holy kiss of the Eucharist does not condemn the passions of his loins and of his heart." And in identifying with the sorrow of the rejected, one still can laugh, with tears in one's eyes. To be a Christian is a "kind of madness," and yet in one's soul is a serenity born of the "sober inebriation of Christ's wine." Worldly struggle in the midst of all this may even lead to "an unheroic ulcer." And yet one knows "the wilderness and Gethsemane, the solitary place before daybreak." And it is in prayer that "he will know the Christ whom he serves with the whip is also he who gently carries the lambs in his bosom." [7]

Perhaps with those who will no longer grant the Church the space in which to expand the height and depth of its liturgy, we might begin with the small and the daily, as we suggested in regard to time. Robert Hamma proposes a "spirituality of space" for the purposes of providing a "landscape of the soul." What he has in mind is to find holiness in the familiar, sufficient to lay claim upon us. A pilgrimage could be to the newspaper stand, where our daily ritual takes us, or to cherished tranquil spots that still get into our bones. Even one's table and dishes can be enhanced so as to give grace to a supper, with only hints disclosed of its meaning. When society deprives one of places that can still claim "holiness," such places will nevertheless appear. Whether or not we acknowledge it, human life has passed beyond the evolutionary point where material can ever again be simply material, and when space is not at least implicitly laced with mystery.[8]

"Holiness" is never absent. What changes is that in which "holiness" is manifested. It occurs whenever space and time intersect for the sake of place, i.e., for "thisness." Viewed functionally, Christianity's "rival church" would seem to be the shopping mall, where Christmas and Easter are played out with a liturgy intensely economic. And a secularized version of monastic hermits could be those in their rooms, immersed in the immensities of cyberspace.

—⊢—

12. Inner and Outer Space

References to "holy space" in the Old Testament move steadily from outer to inner space. Outer forms thus serve as a context for the inner space of "abiding Presence." Christian mystics, such as Abhishikanda, identify this inner space as "the cave of the heart." Many identify the current spiritual quest as an "inward journey." But also crucial is living from the inner outwardly, a living that requires both the clenched fist and the open hand—for waging battle, and for letting go.[9] Existence is an effrontery to nonbeing. Light holds on as long as it can before being claimed by the darkness. Everything that exists deteriorates into nothingness—it cracks, rusts, dissolves, decays, or dies. Nothing can finally withstand the onslaught, and so the inward journey requires intense concentration with the total participation of all our faculties. Ironically, the focus is upon resisting Nothingness, by losing oneself in the All.

In one sense, the inward journey involves taking the outer journey backwards. The more the self becomes mature and centered, to that degree does it crave to lose itself in an Other. It is the return to origins on the chronological plane that brings one to Origins of quite a different sort. It brings us to where Silence speaks boldly, and one dares to deny one's "god" for God's sake. While some contemplatives insist on abandoning all things sensual, at their best, what they are abandoning is their preoccupation with self. And as that fades, the senses return as instruments of holiness. Thus the first barrier to the inward journey is our deep preoccupation with ego.

One day I received a sign that my inner journey was becoming serious. One of my daughters found my three Yale diplomas under the washing machine in the cellar, sopping wet. So as not to offend her, I gave her the proper thanks—and deposited them in a garbage can, next to a wayward pork and beans can. I have often wondered what difference it would have made had Thomas Merton belonged to a monastery where all that was permitted after the title of any book or article was the anonymity of saying, "By a Carthusian Monk." Often it makes a difference when a book is by an author who is writing in order to be known. The person who needs to display diplomas, certificates of scholarship, and awards is a dangerous friend. The morning I discarded the book in which I had meticulously recorded all my published writings was the day I felt a delightful integrity. To be freed of one's "doings" is to be free to "be."

There is a second barrier to the inward journey. It is the tendency we have already mentioned of taking things for granted. The writer of Ecclesiastes knew well this mood. To see one mountain is to see them all. And since every wave is like every other, why waste time sitting and watching? Thus it is that spirituality often begins with an interruption of carefully made plans—an interruption of occupation, health, marriage. Or perhaps it begins upon reaching a certain age. Mine was fifty. For others it might take sixty-five, or even seventy years. The inward journey is most often opened by whatever has the power to flatten our noses up against everything that is terminal. One can no longer take the really important things for granted. Year by year becomes month by month, then day by day.

To begin experiencing things as if for the last time is to begin to experience them as they were the first time. Spirituality comes with the recognition that increasingly everything one experiences is not like every other, but is incredibly special, for it has claims of being the "last time."

Never to see a first snowfall again, or feel the rise and fall of waves, or the astonished eyes of a child is incredible. The irony is that when one experiences anything for the last time, it takes on the incredible "holiness" which it had when one experienced it for the first time. The first time I drank from an icy mountain stream, the first time I fell in love, my first adventure soaring in a sailplane—these are precious times. This is what living is about.

What a shame and what a waste, that so many of us need to reach the end before we can experience the beginning—over and over. What we need is a spirituality that is rooted in, but goes beyond, the symbolic. We need the holy as bathed in Mystery. Last night I swatted a fruit fly between my two hands. It took a certain skill, but I was humbled. What an unspeakable mystery there was in that little fellow—an incredible flier, drawn by instinct beyond imagining, to be dispatched by one who reduced him to being a nuisance. While the special intersections of space and time can evoke a sense of mystery, even to notice them requires an awakening of the senses. I ache knowing how difficult this is in our society—for mysteries once fractured are almost impossible to regain. Tillich claimed that a theonomous situation, *once broken,* can only dissipate into an autonomous state of "self-sufficient finitude," squandering every reservoir of symbol and mystery. Only through crucifixion can the culture and its inhabitants be resurrected to a fresh sensibility for mystery and symbol. The Christian cannot settle for finding symbols here or there. While this is a beginning, nothing else will do than for Mystery to be in, with, for, through, and around, and a craving for, the God who makes Promises. While the Church has a responsibility for the secular world, at this point in time it may best be done through her being restored as a World of Sacraments.

—⊢—

13. The Hermitage of the Heart

Jesus frequently withdrew into the solitude of the hills, especially after a crisis (e.g., the beheading of John the Baptist), before a crucial decision (e.g., choosing the twelve disciples), and when in need of renewal (e.g., being overcome by the crowds). This withdrawal was a need not only for "time alone" but also for a "special place." So clearly was this so that Judas knew exactly where Jesus would be, thus desecrating his sacred place with betrayal. "He came out [from the upper room] and went, *as was his custom*, to the Mount of Olives; and the disciples followed him" (Luke 22:39, emphasis added). In ancient Russia, every village had a hermitage (*poustinia*), available for anyone to use as needed.

Most people today probably regard hermits as bizarre persons who, thankfully, are a phenomenon of the past. The truth is, however, that there are hermits in the forests, and the cities, closer than one might think. *They are necessary*. There have always been persons who by temperament or situation are alone in the midst of people, without understanding why. They might be recluses, in one way or another. But there are others who, living active and rigorous lives in the world, leave it all behind and go into the "desert." Such a hermit vocation is not usually for the young, for it dare not spring from undue idealism or rebellion. There comes a time when one simply becomes tired of pretenses and games. A thirst for integrity takes over, a passion to undertake the austerity of living in simple honesty, without convenience, support, or distraction. This call into solitude is a pilgrimage into darkness and crucifixion, for it annihilates the self one once knew and fostered. It is a lonely path, hidden from the eyes of the world, which neither knows nor cares.

In any case, the world is certain that the hermit is a failure. Free from the lure of possessions, power, and status, the contemplative life has no practical use or purpose. Hermits are pilgrims, dependent on pure faith—faith that this is where God would have them be. To walk into such silence is to be stripped of certainty that one has an answer to anything—until the questions that once plagued the mind nestle in one's soul as friends. One would hardly enter such a valley of shadows willingly. Yet amidst all the options one has, strangely there is no choice. Nothing else matters except to be a person of prayer. And some day, in the gentle quietness, standing among the ashes of dreams and ambition, one may be blessed with the only certitude likely to be given: that to seek is to be sought, and to find is to have been found.

To be drawn into such solitude is really an invitation to share the companionship of God's loneliness—the God emptied in total identification with us—ignored, hidden, forgotten, and profoundly poor. Drawn by such Presence, the hermit lives the joy of simplicity—freed to want nothing more than to grow old loving one's God. Hermits endeavor to remain in the present moment, and since there is neither past nor future for them, the moment is eternal. The soul is the temple of here and now, emptied of everything that might move it from the present as Presence.

And though this complete solitariness is for the few, on behalf of all, each Christian needs a religious sanctuary, a holy space—whether bedroom or basement or a corner somewhere. Family negotiations may permit a note to be hung on the bedroom door for a set amount of time. There, freed of duties for the time being, one can take simple food, a Bible, notebook, and/or crafts—although the ideal is to do nothing. This isn't so weird, if we remember the silent old man on a bench in front of most country stores, silently chewing and whittling. But how does one fill the soul as it becomes a *hermitage of the heart*? One doesn't. The Spirit does—when invited, without distractions.

As over against this life in the soul's solitude, however, stands a certain type of marriage idealized today— of a couple totally open to each other. Absolute openness is to be reserved for God alone, although certain "rooms" may be opened from time to time, but always by invitation only, with shoes removed, because one will be tread-ing on "holy ground."

What "the hermitage of the heart" is about is the establishment of a "soul." I once asked a contemplative how he knew that he was in the presence of God. Without need to think, his response was clear: "A sense of peace." The goal is to have this solitude of inner space for the asking, wherever one is. It comes more easily to children, whose gift of imagination is not yet rendered "wasteful." A child can create holy places anywhere, whatever name they may chose to give them. So the hermit is a model for leaving both memories and plans aside, content to be a guest in the quietly throbbing present.

3

THE WORLD OF SACRAMENT

"Let their table be a trap for them." (Ps. 69:22)

"In the hand of the LORD there is a cup with foaming wine, well mixed." (Ps. 75:8)

"I will offer ... sacrifices with shouts of joy.... I will sing and make melody to the LORD." (Ps. 27:6)

1. Church, Sect, and Mysticism

We have been dealing with the power that symbols, sacramentals, and instruments of holiness have for Christians. Our concern now is to explore the Church as an organic and holy reality, vibrantly alive as the body of Christ, whose very being is as a "World of Sacrament." Jesus' birth, life, death, and resurrection is the primary divine-human event, serving as the center of history, the defining pole of the cosmos, and the ongoing incarnation within every believer. In that sense, we could speak of there being only one sacrament—Jesus Christ. And it follows that, in being the ongoing "body of Christ," the Church is the continuation of this one profound sacrament. The "Church" for Protestants is often seen in more human terms, as an institution and structure for helping individuals with their efforts at Christian living. And yet one of the favorite Protestant hymns begins: "The Church's one foundation is Jesus Christ our Lord. She is a new Creation by water and the Word." And Christ's reason for coming from heaven is precisely because he "sought her to be his holy bride," purchasing her with his own blood, and "for her life he died." No Catholic would have trouble with this understanding.

Furthermore it is possible that Protestant Reformers and the theologians of Vatican II come close in the way they understand the Church functionally. Calvin identified the Church as being present "wherever we see the word of God sincerely preached and heard, wherever we see the sacraments administered according to the institution of Christ."[1] The major Reformers also had a sense for the Catholic view of the Church in terms of "mystery," although they preferred to speak of the Church Visible, and the Church Invisible. And yet since its inception, Protestantism has struggled over some distinctions that Ernst Troelsch rendered classic in 1911 in his *The Social Teachings of the Christian Church.*[2] Here he used the terms *church*, *sect*, and *mysticism* to distinguish ways in which Christians enter into "community." The *Church* is a group that defines itself in broad terms, compromising with the values of society so as to make the Church universally available to all kinds of people in all cultures. "It can afford to ignore the need for subjective holiness, for the sake of the objective treasures of grace and of redemption," Troelsch wrote. Thus in characterizing the Church as holy, this type refers not to individual Christians but to the *sacramental* holiness of the universal Church. The *sect*, in decided contrast, is a voluntary community, bound together by an experience of *"new birth,"* emphasizing individual perfection through discipline. The demands of the Christian faith are difficult, and thus are intended for the "few" rather than for "all," making the sect separate from the society in which it appears. *Mysticism* focuses on the individual, stressing a unique personal inward experience, with a minimum of worship and doctrine. Troelsch asserts that the Catholic Church and the mainline Reformation denominations at their inception shared a common view of the Church. But throughout Protestant history, many sect-type or "free churches" have sprung up, with the number of Christian denominations in this country now approaching three hundred. And the mainline Protestant denominations today are a strange mixture of types, making generalization difficult.

Catholicism, as well as Anglicanism, has found a place for all three approaches within its purview. Traditionally the churches in a diocese are so arranged as to have a parish in each geographic area, thereby striving for a widespread complex of ethnic and class inclusivity. For the broad membership, the "church" type is in evidence, with the sacraments being its primary reason for being. The priest is intended to live a different type of spirituality—one characterized by a discipline similar to that of the "sects," often formed in a seminary related to a monastery, so that priests model the monastic discipline as the spiritual ideal. In addition, there are an incredible number of religious "orders," each one of which has a different flavor, but most entailing sectarian-type vows: of poverty, chastity, and obedience.

The third type, that of "mysticism," characterizes certain orders such as the Trappists, Carthusians, and Carmelites, which emphasize contemplation and a life of prayer as their reason for being. And even further, these orders recognize the validity of certain persons being even more deeply called to a solitary life; these individuals, called *hermits*, usually embrace some form of "mysticism." What monks and hermits do for us is simply to bring awareness of the Mystery into being, continually upheld by the power of God.

—⊢—

2. Protestant Tendencies

Within Protestantism, several tendencies make the denominations complex. First, Protestants tend to understand the Church functionally. The form of the major denominations is of the "church" type, appealing to a broad spectrum of people. And yet, although many churches have abandoned the sectarian emphasis on individual discipline that characterized them at their inception, in practice they function as if they are still a sect—by shunning diversity. Consequently, almost every one of these denominations today has an internal crisis between "liberal" and "conservative" wings. Having little "room" for each other, a split within any of these churches is always possible. A joke has it that one particular Protestant denomination has mastered the art of evangelism. Give one of its local churches a year, and they will have split in two; and each of these, in turn, will have an internal disagreement, and will build another church down the road or in the next community.

Second, at the time of the Reformation, founders such as Luther and Calvin were very close to Catholics in their understanding of the sacramental nature of the Church. The problem was not so much in the understanding but in finding a language that would be able to make misuse and corruption within the church less possible. Yet not even the Protestant Churches themselves were able to agree, and as time passed, varied needs arising from circumstances altered the original "consensus." United Methodism is an interesting example. At its inception, the Methodist Societies were intent upon spiritual renewal within the Anglican Church, and they were able to draw their sacramental life from that Church. When the Revolutionary War was imminent, most Anglican priests returned to England, leaving very few ordained priests to celebrate the Holy Communion. Most Methodist leadership was by itinerant "preachers," who established new churches on circuits. A "District Superintendent" was appointed in order to oversee a given number of churches, visiting each church four times a year ("Quarterly Conferences"). Since these superintendents tended to be the only ordained leaders, even though Holy Communion was very important, it was received only quarterly—when the superintendent visited.

The result of this infrequence over time was a diminishing of the importance of communion. In my childhood, quarterly communion meant that four times a year the church would be less than half full. Presently, however, interest in the Eucharist is increasing, especially among recent seminary graduates. In response, rather than "deprive" any church of the sacrament, a decision was made that any person serving any United Methodist Church could celebrate Holy Communion in that Church, even if she or he was not ordained. Ironically, the "high" doctrine of ordination that once characterized this denomination tends functionally to disappear. A perennial question by candidates for United Methodist ordination is this: "What difference can ordination make? I have already been preaching for several years, and already I can conduct Holy Communion?" And while one is hopeful over the increased interest in the Eucharist, it is at the same time a concern that circumstantial practice is increasing the sacramental chasm.

The third factor making the Protestant view of the Church complex is that interesting events have transpired on the ecumenical plane. One example is the agreement between Lutherans and Catholics on the meaning of justification, the doctrine over which the original Reformation split was largely fought. Ironically, however, even if the top denominational leadership reaches agreement, these results were simply a working through of the theological intent of the original founders of the denomination. But research shows a significant gap between the founders and the current persons in the pew. Not only are the founding beliefs of various denominations little known by many parishioners, but research indicates that denominational preference is no longer a significant factor in the choice of a local congregation. Such dissolution stands in contrast to Luther's deep agony at the Marburg Conference in 1529, where the Reformers attempted to find agreement on the meaning of the Eucharist. As quibbling went on and on, Luther slammed his fist down and wrote with chalk on his desk: *"Hoc Corpus est"* ("This is my body"). For Luther, Jesus himself said these words, and thus it is so! Christ's promise was enough for him—with quibbling about "how" being tiresome and diversionary. But in time, the traditional understanding of the Eucharist, on which Protestants and Catholics were once close, had become so ridiculed in the popular mind that out of these three central Latin words became coined the word "hocus-pocus"— which even today the dictionary defines as trickery, deceit, cheating, and to stupefy with liquor. A once common understanding about the unique centrality of the Eucharist had, in time, become regarded as "magic."

3. Roman Catholic Tendencies

The Catholic Church has traditionally been able to use the authority of priest, bishop, and papacy to stabilize the beliefs of its members. However, the radical changes of Vatican II caused a crisis of belief for many members, much as the Reformation had done four centuries earlier. And while the transition is generally accepted, the present Pope is heralded by conservatives as pulling the Church back toward more "traditional" positions. Yet the more he asserts his magisterial role in an effort to close debate on certain issues, the more he stimulates vigorous discussion and debate. And the more uncompromising the papal statements are about such practices as birth control, ·the more the result is not change but an undercutting of papal credibility. Increasingly papal "declaration" is falling on Catholic ears as though it were merely "suggestion."

In both Catholic and Protestant Churches today, then, a question quickly follows. How is diversity to be creatively affirmed by the Church, while keeping its members grounded in passionate commitment? In the Middle Ages, the Catholic Church, responding to the illiterate condition of the masses, found nonverbal ways to express the faith. Not only were there "miracle plays" in the village square, but liturgy itself became solidified in drama, action, color, smell, gesture, stained glass, music, vestments, and sacred utensils. Learning was through *event*, carefully choreographed—stressing *seeing* far more than *hearing*. Since most people could not read, unified understanding depended upon the authority of the priest, and the Church avoided reading Scripture in any way that would leave its meaning up to the hearer.

Reformers, however, were aware of how nonverbal understandings can give rise to misunderstanding and thus misuse. The cure they attempted was the use of alternative language, opening Scripture in the vernacular to believers, and to public debate. The primary sense became *hearing*, centering in communication through the sermon. The ironic result was that a Catholic emphasis upon nonverbal impact collided with a Protestantism that insisted upon talking about everything—for the Reformers were quite suspicious of communication by *showing*, for it ignited the imagination. The result became graphic some years ago when I attended the village Protestant church in Alsace, France. I could not imagine a colder, drabber gathering, with little more than the pastor's sermon—the service was deprived of all visual expressions and all musical instrumentation. I felt much the same when I saw the "cathedral" in Geneva that Calvin had totally changed so it would be an expression of the Reformation. Reformers were pushed by their own logic to see the sacraments as a supplemental way of proclaiming the gospel. And since the nonverbal was easily subject to error, *preaching became the primary "sacrament"*— with certain sacraments useful but no longer necessary. At its best, Catholic worship is a matter of "show and tell." Protestant worship, in turn, involves "tell and show."

It is hard to overemphasize the impact of Vatican II on Catholicism. The central momentum was that beliefs and practices would be subjected to critique in the light of Scripture and the early Church. As a result, at Sunday masses preaching (giving a homily) is now mandatory; lay study of freshly translated Scripture is encouraged; more democratic forms of administration are being implemented; and spiritual practices are no longer just for priests and monks, but are encouraged for the faithful. Lay lectors now read the Scripture at Mass (the priest reads the Gospel); the cup is offered to parishioners; lay Eucharistic ministers can offer the elements and have entry to the tabernacle; the Bible has been translated into several versions, and lay folks are encouraged to own one. In addition, encouragement of spiritual practices for all has led to daily Eucharist and a new version of the Breviary for lay use. Above all, the liturgy is totally in the vernacular—to *see* and to *hear*.

The total result is that Vatican II has done for Catholicism much of what the Reformation had proposed. The functioning of the laity within liturgy recalls the Reformation encouragement of the "priesthood of all believers." Likewise, Luther's translation of the monastic tradition into a new organic image of the Christian family finds expression in Vatican II's emphasis on the spiritual family. Most Catholic theologians agree that if the equivalent of Vatican II had happened in the sixteenth century, the split between these two great Churches would never have been necessary. To help Catholics understand these huge changes, then, particularly in the liturgy and the issue of diversity, the focus of the Church needs increasingly to be on a deeply informed and educated laity.

—┬—

4. Reaching Toward Each Other

Simultaneously with the rapid changes within Catholicism, there has been a growing restlessness in Protestantism about verbosity. Its tradition had given rise to the incessant need to talk about everything, accompanied by a mistrust of symbol, imaging, and physical participation. The price for needing to explain everything verbally, however, was an undercutting of any real need for the nonverbal. Music by choirs was disengaged from its original function of serving the liturgy, resulting in its becoming more of a performance, even eliciting applause from the congregation. Recently, however, I detect a growing hunger for something more. The term for this "more" is "spirituality," a term that was not much used until Vatican II. Whatever else it might mean, it represents an effort to get beyond an intellectual approach to Christianity for the sake of "living" the faith, and a movement beyond a frenetic doing toward the quality of "being."

Also growing has been an interest in more frequent Eucharist—first within the clergy and now among the laity. Likewise, conservative and fundamentalist churches are using rock music and alternative means of communication to create worship as "event," inviting physical participation.

New hymnals have been produced by most denominations, with the result that Catholic hymns appear intermixed with Protestant ones. Differences between the two hymnologies have turned out to be inconsequential. Symbolic of this is that I have yet to see a recent Catholic hymnal that does not include Luther's "A Mighty Fortress Is Our God."

The liturgy of most mainline Protestant denominations has been redone, making appreciative use of the excellent work by Catholic scholars in revising the Mass after Vatican II. As a result, the liturgy of the Roman Catholic Mass and that of Holy Communion for mainline Protestant churches is becoming creatively similar. And these ecumenical tendencies, in turn, flow back upon the denominations to give a new sense of the importance of the tradition flowing from their inception. The great Eucharistic hymns of Charles and John Wesley, which were largely absent from the former United Methodist hymnal, have reappeared, within a deepened appreciation for the Eucharist as sacrifice.

In the spiritual direction I do with Protestant pastors, I find it interesting how many find spiritual strength in quietly attending a daily Roman Catholic Mass in a neighboring town where they can be anonymous. Likewise, the guest wings of Catholic monasteries are usually full, often with as many Protestants as Catholics. Twice a year, I teach an action/reflection class for United Methodist seminarians called "Monastic Spirituality Immersion." It takes place at my monastery, where for a week they *are* monks. Instead of being offended by such incidentals as rosaries and statues of the Virgin Mary, the students have been able to look beyond these particulars to the spiritual foundation being lived by monks. Almost to a person, the consensus that emerges is that Trappist monasteries are to Catholicism what Wesley's Methodist Societies were to Anglicanism—powerful attempts at spiritual renewal.

As the above examples would indicate, these two great expressions of Christianity—Catholic and Protestant—seem to be moving toward each other in the time and space of their worlds. Catholics are moving toward a more liberal understanding of tradition, while Protestants are recovering nonverbal ways of being spiritually fed, searching their own traditions for direction. This is encouraging, but I am anxious lest we pass each other in the night. Significantly, each tradition is recognizing a yearning for what the other has. Yet it is crucial that in the process neither forfeit the richness of its own tradition. Christian groups tend to be right in what they affirm, and weakest in what they deny. If we are to restore the Church to its full stature, evident in the rich tapestries that were present before the unfortunate Protestant-Catholic split, it is crucial that we realize how much we need each other. The Church's renewal requires freedom *and* authority, mystery *and* understanding, prose *and* poetry, altar *and* pulpit, sacrament *and* sacramental, silence *and* sound, spirituality *and* social justice, liturgy *and* spontaneity, doing *and* being. To hold these together is finally to know that Catholicism as the Church of Peter, and Protestantism as the Church of Paul, find their completion in each other. It is finally to affirm that the priest at the altar is "standing in" for Jesus himself. And likewise, when the pastor reads the Gospel, it is Jesus Christ himself who is proclaiming the Good News. Appropriately, in the Byzantine Rite the "Presider" emerges twice through the great door—to proclaim the Gospel, and to bring communion to the people.

—┬—

5. Spirituality and Creation

The emergence of interest in spirituality over the past decade is resulting in a new "ecumenism of experience." To read in this "field" is to become oblivious of denominational difference. Writers such as Thomas Merton, Tilden Edwards, Henri Nouwen, Annie Dillard, Kathleen Norris, Matthew Fox, Madeleine L'Engle, Sam Keen, and Dag Hammarskjöld—these are meant for all of us, with little need to separate them along denominational lines.

This helps us identify the point that we have reached in our exploration. The cultural emptiness and the secular searching that characterizes our time is a craving to discover more than isolated supplements for modifying a drab existence. Known or unknown, our modern restlessness can scarcely be sated with anything less than what the Church can provide. And yet the Church, almost overcome by the secular vantage, must once again become what she is. By her nature, she is an organic, living "World" that requires one to undergo careful formation in order to become a serious resident. In the struggles and division, arguments and struggles, Catholic and Protestant, we have tended to lose the reality of the very Church that has supposedly been our concern.

Therefore we need to apprehend more fully the Church's essential nature as the "World of Sacrament." In the first book, we explored the Triduum (the holiest days of the Christian year, from Maundy/Holy Thursday through Easter) as an experience of this essence. It is a living event with the power to transform. From the Triduum as the "primal event" of Christianity, the "primal image" that emerges is the Eucharist. Yet how can all of this be held together as a genuine alternative to the secular world in which we exist—on the edge of losing ourselves? The transformation centers in recognizing the Church of Jesus Christ as an authentic *alternative* World. As its structure, the sacraments appear at each major fork within the human pilgrimage, evoking expansive sacramentals. It is deep with symbol, in which all that *is* is more than it appears to be. There are special spaces of holiness, with a richness from the various dimensions of time, formed and rehearsed within the Body that is home.

For this to be possible, we must resist certain tendencies often characteristic of spirituality, or what Troelsch identifies as mysticism. First, the mystic approach tends to become separate from the Church unless it is firmly contained within the corporate arena of monasticism. Secondly, mysticism tends to deny the reality of that which is of the earth, the flesh, the joyous substance of existence. It is crucial that Christians resist any spirituality that denounces the time-space called creation—seeing it as a threat, a temptation, or a distraction from spiritual life. Even in some classical spiritualities, we find tendencies that can be interpreted as world-rejecting. The Christian must interpret these as affirming detachment *from* the world as a way of becoming free *with* and *for* the world. The Christian faith must be firm that the dwelling place of God is *with* creation, just as the dwelling place of humanity is *in* God. To be able to live within this understanding, our senses need to be purged—for the modern world has dulled them, by excesses of noise, sights, and images flooding our senses into insensitivity.

In a significant sense, "reality" is in the eye of the beholder. For the Christian, this means an eye honed by the Eucharistic community. What we can see from that vantage point is the beveled edge between Spirit and World as the frontier of ongoing meaning. The brilliant early Lutheran thinker Jakob Boehme rightfully identified the goal of redemption as healing the apparent cleavage between Spirit and nature everywhere. Where this occurs, we have beauty—standing as judgment on our society that humiliates beauty. For the Christian to befriend beauty means to give form to the chaotic and birth to the potential, in each case rendering one's living spot a place of foretaste. It is immersion in the Eucharist that can make adoration and awe basic attitudes of one's soul, attitudes that alone know what it means to find delight in the splendor and beauty of God. Unless one's soul learns early how to sing with one's body, work is a coarse fate.

John Cassian's classic list identified for the Church eight primary expressions of sin: gluttony, fornication, material security, anger, melancholy (to feel sorry for one-self), boredom or monotony, vainglory, and pride.[3] Yet there can be no healing for any of these without delving into the deeper cause. In the first book, we identified the "primal image" of Christianity as the Eucharistic gesture—the lifting up of the consecrated elements of bread and wine to God. In truth, *no one can escape having some sort of "primal image,"* as we saw in insisting that every person has an "ultimate concern" or primary object of trust. Whatever one's primal image, it functions as the glasses through which one sees everything. And it follows that if this image is unclear, self-serving, or parochial, one's behavior, which flows directly from it, will be flawed. *The resident cause of sin, then, is the primal image that structures one's perspective on everything.* Thus to change one's fundamental behavior requires a change of primal image. Actually, conversion means the death of one's primal image, and the rebirth of quite another one. Such replacement is the meaning of redemption.

The sociologist Margaret Meade is referring to a primal image when she affirms that each person constructs the past from the vantage of the present, in light of the future. Even the past can be redeemed in the light of the primal image in the present. Thus it is not yet clear if the

Civil War was a tragedy or a liberating cause, or both—for the results are still operative in the present. Likewise whether the history of the human race is one of pathos, or unimaginable glory, is yet to be decided. And the Christian, living liturgically within the Triduum, claims the future in terms of God's promise—trusting as we raise the "cup of eternal salvation." Thus we celebrate history by reading it backward and then forward—from the Christ event as center.

Unlike what we often hear, then, the Christian is not really striving to be "perfect" in behavior—but striving first for a consistent faithfulness to one's primal image. Thus the parables are to Jesus what the sacraments are to the Church. It is through participation in the sacraments that we learn how to live the parables. *The sin of our age emerges from the primal image by which it constructs its world.* Deadly is the gaze of suburbia, where the final spontaneity to eliminate is a wayward blade of crabgrass—exercising therein an "ultimate" expression of control. Without *wilderness*, as we insisted in the first book, the soul withers. Yet it is frightening to see the way we participate in the set-aside areas that we do have. Supposed wilderness areas are being distorted as tourists sift the "wilds" for the photogenic. Drive your RV to our doorstep. And once there, the scenery competes with the TV screen, and buying postal cards takes precedence over participation. Dune buggies, ATVs, and snowmobiles thunder through the wilderness we once knew in silence.

I live in a cedar forest, at the edge of a placid lake, in a hermitage I constructed from wood salvaged from tenement houses in the inner city where I lived. Its main supports were redeemed from a former house of prostitution. I live in beauty, tranquillity, and joy—with breezes through the summer, snow bending the tree boughs in winter, and greenness everywhere in the spring. And now in the fall, multicolored leaves flutter down to adorn the forest floor. It feels strange when most of the persons who buy land in this area to build houses cut down the forest so that they can plant and fertilize a large section into suburban grass, which needs to be cut on a riding mower. All depends on the image through which we see everything. I am envious of the medieval period when the prevailing perspective could discern symbolic meaning in so much of life. While Genesis insists that we are called to be gardeners of creation, this is possible only after conversion of one's primal perspective. Thus when one is centered within the Church, the time and space of our location becomes surrounded by sacramental invitations. As a result even our "possessions" we possess as if not. Everything, simply everything, depends on the primal image that structures one's perspective.

—⊢—

6. The Citadel of Presence

Dietrich Bonhoeffer was perceptive when he insisted that the Christian never sees *anything* nakedly or directly. The mark of conversion, then, is one's ability to see other human beings as literally the ones for whom Christ died. In Christ we have been claimed, not *away from* our humanness, but claimed and ignited gloriously *in* our humanness. Thus the *created universe is the temple of God, with the Church her soul, where on her altar all of creation is lifted up and blessed. And when it is returned, the world glows with promise, and all of us are sanctified as secular priests to render sacred that which we see and touch and smell and hear and taste.* Jesus Christ is *the* Sacrament, and it follows that as the "Body of Christ," the Church is the *ongoing* sacrament in the world. The heart of the Church, her very soul, is the Sacramental Word and the Word as Sacraments—as the "ongoingness" of her incarnate life. In trusting the promises of Christ who continues to act within her, the Church is a living foretaste for all of the new heaven and the new earth. Looking out from within her, one perceives the vision of the sacredness of even the secular. We can see God in the created things around us—birds, flowers, and hills—the things that Jesus loved.

Yet it is the sacraments that prepare us for even more than seeing. Through them we begin to experience "Real Presence." And as these sacraments converge, the Church as the "World of Sacrament" so awakens our senses that the Church is experienced as a *Citadel of Presence.* The

Church is a living Mystery, standing in the midst of a world created by an unfathomable Secret. Sacramentally surrounded, we understand what it means for the event of Jesus Christ to be the declarative Incarnation, so that crucifixion, resurrection, and ascension are the marks of Christ's *Presence*. We can understand the Protestant practice of locking the Church door when the congregation is not there, for the congregation *is* the Church. Without them, the building is a vacant meeting place. And yet there is a difference when the Catholic Church endeavors to remain open, that people may come into a Presence, into an intensity of the holy, with an imaginative evocation of the sacramental—a "Citadel of Presence." This helps to make clear the Catholic's sense of emptiness, abandonment, barrenness, and vacancy one experiences on Good Friday and Holy Saturday within the Church as the Absence of the Presence. In contrast stands the beauty and power of the liturgy, especially on feast days, where in that very time and space one is bathed in a profound sense of glory, majesty, and beauty. There, to be experienced in a profound sense, is a festival of the heavenly country. And it remains in that place, as a sense of joy in Presence.

—⊢—

7. The Primal Image

The great novelist William Faulkner was once asked why there were so few individuals in his novels who experienced redemption. His answer was quick: "I have always regarded God as being in the wholesale business, not the retail trade." However one might respond to this, for the Christian truth comes first and foremost through participating in the meaning of the Whole, in its broadest and deepest sense. Put another way, the liturgy of the Church is the distillation of the primal rhythm that structures time and space.

Many of the fairy tales on which we were raised as children are based on a rhythm inherently Christian. It is the rhythm not of seeking and finding, but of seeking, *losing*, and finding. Viewed in terms of personal pilgrimage, God is in the lost-and-found business. Viewed in terms of the cosmic perspective, we have seen time as variations on a triad: that of promise, gift, and response, or the anatomy of creation, fall, and redemption. These, often together, form the perspective through which Christian eyes perceive reality, distilled in the event of Eucharist—as Word and Table. With this perspective in place, the Christian is open to such a "scientific" macrocosmic description as the "Big Bang Theory." This description of origins can be imaged by the Christian as centered in an Infinite Intense Source, which is continually being born as Infinite Expansion, returning as self-conscious Enrichment through humanity as a self-conscious center. Stated scripturally, it is in Christ and through him that all things were made (John 1:3); and it is to him that all things return, for in him all things exist (Col.1:16-17). Every day the primal movement, miniscule or gigantic, is from crucifixion through resurrection to ascension—with the Eucharist functioning as the primal drama through which this movement is observed symbolically as an offering up of creation into the becoming of God. *The Divine promise is that space and time, given as gift and calling, pass into eternity as increase.* Expressed more personally, the works of our hands, the dreaming of our minds, the cravings of our souls—these are never lost, but make fuller the *Pleroma,* the fullness of the Whole. There is a difference between something that *may* be an addition, and that which is given in Christ's name *as* an addition. It is different to give a cup of cold water to one who is thirsty, and to give it in Christ's name.

The creation, not yet finished, is our arena in which we live as co-creators with God—in the name of the new heaven and new earth. We are called to alternate between the vigor of work and the intensity of contemplation, both of them foretastes of God's Sabbath of infinite stillness and bliss. To receive all as gift, and to return all as increase, is to vanish into God. And so daily we plunge into the mystery of God, losing ourselves in the Silent Center, arising in foretaste of Eternity. The supreme declaration of the Christian needs to remain clear—that God, in whom we live and move and have our being, has "pitched God's tent among us." When and where these truths intersect, there the melodies of praise and glory abound—in the ebb and flow of hours and seasons, places and things.

Redeemed eyes can behold the rich shades of autumn that give way to the sacred stillness of winter in holy waiting. And the silence shouts in shameless abandon, as all that *is* takes root in the promise called spring.

Since New Testament times, the primal image that transforms our senses has been called the Triduum ("three days") —the wholesale perspective, the Event of events. If a stranger wanted to participate in the meaning of the Christian faith, the best way to do so would be to experience the Church's worship from Maundy Thursday through Easter.[4] In living these three days, one can best experience from within the meaning of the All. While words can try to point toward its meaning, they function primarily as poetry, just as the Triduum itself is a composite of drama—light and dark, sound and silence, nothingness and everything, action and gesture. It draws us in, so that our participation is a living and vital one. This liturgy is the definitive "dress rehearsal" in terms of which one is enabled to recognize this ongoing pattern in the reinforcing of the Church's sacraments and sacramentals. With our senses honed by the Triduum, the world, in turn, is increasingly capable of being seen sacramentally.

Kenneth Leech identifies creation as the first fundamental sacrament, the second being the Incarnate Christ, and the third being the Church, derived from Christ as the extension of his Incarnation. This is simply a way of extracting the faith logic implied in the Triduum—the original context that rendered the Church a World of Sacrament. So rich is the liturgy of that event that phrases and gestures articulate in imagery the meaning of all time and space. For example, early in the liturgy the priest pours a small amount of water into the wine to be blessed, doing so with these words: "By the mystery of this water and wine, may we come to share in the Divinity of Christ who humbled himself to share in our humanity." One simple pouring, several poignant words—and we are at the center of the meaning of the Cosmic Enterprise. The Eucharist is the distillation of Christian revelation. Eucharist as the anatomy of the Christ event is the choreography of God. Incarnation, crucifixion, resurrection, ascension, pente-cost—it is all there, as past becomes radiantly present. It follows that Christian existence is Eucharistic existence.

—⊢—

8. Sacrament and Sacramental Revisited

In summarizing what we have been exploring, we recall the distinction between *sacrament* and *sacramental,* a dis-tinction credited to Peter Lombard in the twelfth century. The meaning of *sacrament* comes from the Greek word *mysterion,* meaning "the secret thoughts of God." Likewise, the word *sacrament* is derived from the Latin word *Sacramentum,* meaning "an oath of allegiance," or a vow to keep a promise. Thus a *sacrament is the Mystery in which the Divine Promise faithfully intersects with our need.* Appropriately these promised sacraments intersect with the crises or decision points of life: birth, identity, calling, sin, sexuality, sickness, nourishment—and death. A sacrament both contains and conveys grace. It involves matter, form, priest, and intent; while the recipients need to be of proper "disposition." Traditionally for the Catholic, four of the sacraments (Eucharist, Reconciliation, Marriage, Healing) are regarded as indelible, with three (Baptism, Confirmation, and Ordination) also impressing a character upon the receiver. Thus sacraments are not only effective, but are inclined to be permanent. Yet while they are effective, they may not be fruitful. These seven sacraments are the Catholic Church's redemptive response to the seven deadly sins. They are special intersections whereby God touches our lives in a unique way, *through promised channels*, and they operate by invitation only. Sacraments do not place limits upon God, but are the gift of guarantee for us. The emphasis is *"ex opere operator"*—that what happens is an act of God, not our doing. And as long as the conditions for validity are observed, and the subject places no obstacles in the way, the sacrament is efficacious. In no way are sacraments to be understood as something God *had* to do. Instead, they are rooted in God's free, spontaneous, and gracious choice. Thus any attempt to understand sacraments in any fashion other than as God's Promises is futile.

On the other hand, a *sacramental* for the Christian is an object and/or gesture that evokes in us an awareness of what is so, independent of the sacramental itself. Thus, while sacraments are signs and causes, sacramentals are signs but not causes. And while one's disposition is a factor in both, it is especially so with sacramentals, because what we bring to them is primary. One dare not underestimate the power of sacramentals. Sons and daughters think they are helping their aging parents by moving them to a retirement village, attractive in its curved walkways and manicured lawns. Often a parent dies soon thereafter. Whatever one may say about the condition of the home in which they have lived for many years, *it is a sacramental*—in whole, and in each of its parts. Even dents and scratches have stories to tell, so that the walls are adorned with sorrows and joys of a remembered past. Even a pair of Mickey Mouse ears can be a "treasure" of *anamnesis*. Suppose you lose your wedding ring, and call your spouse, telling him or her not to worry because you'll drop by Wal–Mart and get a new one. If you do, forget it—your marriage is over. We tend to overlook this power of sacramentals, or reduce it to "sentimentality."

Protestants often deny that they have any sacramentals. When I hear ministers say that, I dare them to worship next Sunday without a bulletin, or to move the American flag, or to put in storage the sports trophies in the fellowship hall—or even to borrow, unasked, a pot from the Church kitchen. To miss a birthday or anniversary of a loved one can inflict deep pain. Sacramental times arise everywhere with every situation and every person. It is part and parcel of being human. It involves a favorite ornament for the Christmas tree or an Easter basket one has had since a child. There is not one of us who doesn't have many things "we don't need," and yet to throw away many of these is to lose something "precious." Almost everything we do is an unspoken liturgy—the order in which one dresses; the "usual" for breakfast; the same pew at church. All of us have "favorite everythings" and are creatures of "habit." God uses these sacramentals to lure us into thought, desire, and action. But it is important to understand that these are reminders, not sacraments. And from time to time, one needs to walk through one's home or work place, recalling, reclaiming, and recommissioning each sacramental. They fluctuate in the depth of meaning—as new and old ones compete for one's attention.

In defining terms in this traditional way, it would appear that some of the sacraments in various denominations have unknowingly become "sacramentals." This is a decided tendency within Protestantism. The Catholic Church, in contrast, needs to be sensitive to the tendency for certain sacramentals to approach idolatry—attributing to them the power of a sacrament. Put succinctly, *a sacramental evokes awareness, while a sacrament effects that which it signifies.*

—┬—

9. Transubstantiation:
The What and How of Sacraments

The understanding of the Eucharist by Luther, Calvin, and later John Wesley, shows a remarkably "high church" character. But in attempting to abolish excesses in the use of sacraments, they tried to use language that would be distinct from the traditional Catholic language. This was especially so with such key words as "transubstantiation." But in the absence of a commonly accepted alternative theological language, confusion and misunderstanding about sacraments was inevitable.

My graduate training was possibly the best available anywhere. In studying the sacraments, however, I was taught that the early Church understood Holy Communion as a powerful practice for remembering the death of Christ on one's behalf. There was no agreement on how many sacraments there were. Augustine kept the issue open by declaring there to be an indefinite number, for the Spirit can act through whatever and whenever it chooses. In AD 836, a monk named Radbert proposed an understanding of the Lord's Supper that anticipated the later doctrine of "transubstantiation." His view was that when the priest said the "words of institution," God miraculously acted upon the bread and wine so that they became the actual body and blood of the Lord, even though their appearance remained unaltered. At first this position did not evoke a wide interest. Around the same time, a monk named Ratramnus took a contrary position. He held that the Lord's presence in the elements is spiritual, with neither physical nor metaphysical alteration occurring. Neither did this view receive wide acceptance or official approval. Yet within two centuries, the "transubstantiation" view of Radbert had become so popular in usage that when Berengarius of Tours reaffirmed the more liberal view of Ratramnus, there was such a public uprising that he was forced to recant. The conclusion that I was taught was that the "crude view of the peasants" rather than the "clear thinking of theologians" had won the day. Thus, Protestantism usually asserts that the understanding of the Eucharist *evolved* into "transubstantiation" during medieval times.

My own pilgrimage, however, involved discerning the degree to which a "high church" approach to sacrament is rooted in scripture itself. At least two sacraments have their clear origin in the Gospel accounts of Jesus. Prior to the Ascension, Jesus clearly gives a mandate for Baptism: "All authority in heaven and on earth has been given to me. Go therefore and make disciples of all nations, baptizing them in the name of the Father and of the Son and of the Holy Spirit..." (Matt. 28:18-19). Even without counting the Eucharistic foreshadowings in the feeding of the crowds and the changing of water into wine, the Gospel writers give a significantly disproportionate space to the last days of Jesus' life, especially the Last Supper. In Matthew, nearly one-third of the Gospel is concerned with the final days, and in John it is almost one-half. Clearly the Triduum is the central image within the Gospels, shaping the disciples and the early Church with its indelible importance. Part of its uniqueness is that its mandate is internal to the event itself. For example, after Jesus' death, a soldier pierced his side with a spear, and out came water and blood—bearing witness to the primacy of the sacraments of Baptism and the Eucharist.

Further, Oscar Cullmann, a well-received Protestant biblical scholar, makes a persuasive case that the Resurrection is directly intertwined with the Eucharist. In Luke's account of the distressed disciples on the road to Emmaus, the disciples did not recognize the resurrected Jesus by his appearance, but through one special act. His identity was disclosed when he did what he had done so often when physically present with the disciples: "He had been made known to them in the breaking of the bread" (Luke 24:35). What's more, the account of this event clearly includes the four "acts" that to this day have been understood as structuring the Eucharist. "When he was at the table with them, he *took* the bread, *blessed* and *broke* it, and *gave* it to them" (Luke 24:30, emphasis added). The consequence? "Their eyes were opened, and they recognized him; and he vanished out of their sight" (Luke 24:31). Furthermore, almost every time that the resurrected Jesus appears to the disciples, they are eating and drinking together, often in the same upper room where the Last Supper took place. On one occasion, Jesus himself sets the food before them. "Jesus came and took the bread and gave it to them" (John 21:13).

Paul apparently knew nothing about an empty tomb. His conversion came through a powerful resurrection appearance on the road to Damascus. Paul experienced a dazzling light that rendered him blind. For the disciples, the resurrected Jesus had a physical appearance different from the one they knew before the crucifixion, for they did not recognize him by sight. Mary Magdalene did not recognize him, and because of his nature, Jesus would not permit her to touch him (John 20:17). In these appearances, he was not literally physical for he passed through walls and closed doors.

Jesus had promised that he would never leave his disciples orphans, but would send the Holy Spirit. The Gospel writers portray this coming in different ways. In fact, the "Holy Spirit" and the "Spirit of Christ" were interchangeable expressions. There is good reason to hold that from the earliest time, the disciples' experience of the Spirit of Christ was so present in the breaking of bread that it was as if the very Christ who had been with them in the flesh, when they broke bread, was now with them as a "Real Presence." So powerful was this Real Presence that nothing less than words such as "resurrection" could express or account for what happened—and continued to happen. When the early Christian community "assembled 'to break bread,' they knew that the Risen One would reveal his presence in a manner less visible but no less real than previously." [5]

This movement from remembrance of the "Last Supper" to the ongoing Real Presence of Christ in the "Lord's Supper" marks a crucial transition. The Church grasped early that what Christ intended by what they had assumed to be the *last* supper, was in fact the *first* supper of an endless stream of resurrection banquets. His promise was of "Real Presence" as his ongoing Incarnation within the Church—as his body broken and resurrected. This understanding is clearly affirmed in Peter's sermon. "God raised him on the third day and allowed him to appear, not to all the people but to us who were chosen by God as witnesses, and who ate and drank with him after he rose from the dead" (Acts 10:40-41).

It is impossible to dismiss that this "high" view of the Eucharist appears in the Gospel of John, written around AD 90. By that time the Eucharist had begun to solidify in practice and meaning. John's Gospel is fully sacramental, using as symbols such daily materials as bread, living water, inextinguishable light, mountains, life, feasts, wine, grain, word, shepherd, sheep, healing, and doorways. Graphic images appear throughout—such as the marriage feast of the Lamb, the pool of water stirred by the Spirit, the Temple as the Body of Christ, and rebirth through water and the Spirit. Christ is the Word made flesh in such a way that we can behold his glory. Throughout, Jesus acts and speaks in signs, requiring sacramental eyes in order to see.

Thus John's Gospel is an invitation into the "World of Sacrament." This is powerfully declared in John 6 with the image of the New Passover, in which Jesus provides bread for over five thousand persons—the bread being distributed only after "he had given thanks" (John 6:11). It was not only sufficient for all to eat, but twelve baskets of fragments remained, symbolizing bread for the twelve disciples whose mission it would be to feed others in turn. Afterward, people came "looking for Jesus," because they ate their "fill of the loaves" (John 6:24, 26). Jesus then introduces them to the "food that endures for eternal life" (v. 27). What he gives is the "true bread from heaven," which "gives life to the world" (vv. 32, 33). Their reply is direct, even eager: "Sir, give us this bread always" (v. 34). Jesus regards them as ready to hear the deep connection to himself: "I am the bread of life. Whoever comes to me will never be hungry, and whoever believes in me will never be thirsty" (v. 35). Using the image of manna given by God in the wilderness, he states quite blatantly that the bread one must eat "for the life of the world is *my flesh*" (v. 51, emphasis added). When these words elicit skepticism from many hearers, he does not back away, or try again with more "spiritual" language. Instead, his response is a declaration, expressed in as dogmatic and negative a way as one could imagine: "Very truly, I tell you, unless you eat the flesh of the Son of Man and drink his blood, you have no life in you" (v. 53). Lest he be misunderstood, he repeats it yet again, this time positively: "Those who eat my flesh and drink my blood have eternal life" (v. 54). He repeats it still another time, even more emphatically, if that is possible: "For my flesh is true food and my blood is true drink" (v. 55). Then in an even deeper fashion, he declares that this Eucharistic act is the expression of his ongoing incarnation: "Those who eat my flesh and drink my blood abide in me, and I in them" (v. 56). Again: "Whoever eats me will live because of me" (v. 57). And again: "The one who eats this bread will live forever" (v. 58).

There is no way that these thoroughly Eucharistic declarations are somehow a misunderstanding, or that the listeners were wrong in understanding these teachings in a somewhat literal fashion. The repetition makes only too clear that Jesus indeed meant what he said, over and over. And because he repeated these words enough times for the listeners to understand that he meant what he said in all seriousness, the sacred moment was at hand. "When many of his disciples heard it, they said, 'This teaching is difficult; who can accept it?'" (v. 60). Jesus knew very well what they were murmuring, and so it is significant that he again refused to make his declarations less difficult by trying a more symbolic or "spiritual" rendition. In fact, the opposite was the case. He insisted, even more strongly, on what he had said, using their disbelief as the context for declaring that to believe in what we are calling the "World of Sacrament" as the nature of Christian life is not within a person's own power. So difficult to believe is this central issue, *that the faith to do so is the supreme gift of God.*

Then comes the great division. "Because of this many of his disciples turned back and no longer went about with him" (John 6:66). And "For not even his brothers believed in him" (John 7:5). In fact, even the many who were amazed at his learning responded in anger, "You have a demon!" (John 7:20). And as the desire to kill him grew, even then he did not back down one bit. "On the last day of the festival, the great day, while Jesus was standing there, he cried out, 'Let anyone who is thirsty come to me, and let the one who believes in me drink'" (John 7:37–38). With arrest becoming imminent, his imagery began moving from bread to drink, as with spilt blood. He emphasized that this would happen soon. His healings became increasingly healing of eyes—for the listeners, in refusing to believe what he plainly stated, were not simply blind physically, but *sacramentally*. One blind man, for this healing, was immersed in a particular pool, suggesting Baptism. "He went and washed and came back able to see" (John 9:7). The testimony of this restored man is posed as a creed for all baptized believers: "One thing I do know, that though I was blind, now I see" (John 9:25). At this point Jesus' mission can be recognized. He came "so that those who do not see may see, and that those who do see may become blind" (John 9:39). John's Gospel is in many ways a portrait of feast after feast at which Jesus' words give meaning to the event (John 10:22f.; 12:1f.). In this context, Martha's confession is interesting, speaking of Christ as the One "*coming* into the world" (John 11:27, emphasis added). From this point on, image after image discloses this "coming" as the ongoing incarnation in those who believe.

The Letter to the Hebrews, written earlier than the Gospel of John (probably around AD 70), reinforces the centrality of the Eucharist which we find in the Gospel of John. But Hebrews does so more by emphasizing Jesus Christ as the supreme *priest*. "He had to be like his brothers and sisters in every respect, so that he might be a merciful and faithful high priest" (Heb. 2:17). He is a cosmic and eternal priest, for "we have a great high priest who has passed through the heavens, Jesus, the Son of God" (4:14) Yet this one, who was God, was also like us in every way, "who in every respect has been tested as we are" (4:15). Therefore as high priest he makes the offering unto God, but as human it is his very humanity that is offered. So Jesus Christ is both offerer and offered; he offers himself. In his life as human, it was his "loud cries and tears" that were offered up, and therefore he was "designated by God a high priest according to the order of Melchizedek" (5:7, 10). Melchizedek is the mysterious figure who, in seeing Abram, "brought out bread and wine; he was priest of God Most High" (Gen. 14:18). But even more, of Christ God has said, "You are a priest forever" (Heb. 5:6; 6:20; 7:3; 7:17; 7:21). And those for whom Christ renders the sacrifice are those who "inherit the promises" (Heb. 6:12, 15, 17). The writer keeps piling up grand images of Christ's priesthood. Not only is he "a high priest forever" (6:20), but he "enters into the inner shrine behind the curtain" (6:19), which is the Holy of Holies where God alone dwells. Christ's priesthood is "not through a legal requirement concerning physical descent, but through the power of an indestructible life" (7:16). Thus "he holds his priesthood permanently, because he continues forever" (7:24)—"as high priest, holy, blameless, undefiled, separated from sinners, exalted above the heavens" (7:26). It is Christ himself who is offered up as host, whereby "this he did once for all when he offered himself" (7:27). As "a minister in the sanctuary and the true tent," this priest has "something to offer"—namely, himself (8:2-3). The writer then describes in detail the objects needed in the Hebraic sacrifices. They correspond very closely to those of the Catholic, Orthodox, and Anglican Churches—a lampstand, a table with the "bread of the Presence" as the

"Holy Place," behind which is the Holy of Holies, an altar of incense, and the Ark with precious objects (9:2-5). It was Christ who entered such a place, and gained redemption by the shedding of "his own blood" (9:12).

After summarizing the whole of Israel's history, the author draws the clear conclusion that they all lived by *trusting the promise* of what God would do. On this faith we all must stand—*trusting that the God who promises is faithful*. And we, who have been promised the Kingdom, need now to "offer to God an acceptable worship with reverence and awe" (12:28). And yet, while Jesus was sacrificed and raised for us once and for all, it is also ongoing—then, when, and now. We know this "Jesus Christ is the same yesterday and today and forever" (13:8). "Through him, then, let us continually offer a sacrifice of praise to God" (13:25). Finally, the overall image that brings the argument to a crescendo tastes deeply of the Orthodox celebration of Eucharist. "Since we are surrounded by so great a cloud of witnesses, let us also lay aside every weight and the sin that clings so closely, and let us run with perseverance the race that is set before us, looking to Jesus the pioneer and perfecter of our faith, who for the sake of the joy that was set before him endured the cross, disregarding its shame, and has taken his seat at the right hand of the throne of God" (12:1-2).

From such Scripture to the early Church is a short step. In Justin Martyr we find the same understanding (ca. AD 153). Writing to Emperor Antoninus Pius he wishes to make clear the heart of the Eucharist. "We do not receive these things as common bread or common drink.... the food over which thanks have been given by the word of prayer comes from him—that food from which our blood and our flesh are by assimilation nourished—is both the flesh and blood of that Jesus Christ who was made flesh." [6]

In order to understand the nature and meaning of the various sacraments, we need to take one step further in understanding the Eucharist as the Church's central sacrament. Throughout Church history one can note an unfortunate and recurring tendency: the tendency to confuse, and sometimes make identical, the *what* and the *how* of a doctrine or sacrament. Or again, often there is a mixing of *confession* and the attempt to explore the *probability* of some event. One example is the confession of God as incarnate within the event called Jesus Christ. That confession is central to the Christian faith. But there is a different question: "How can that be?" For the believer, the confessional answer would be that Scripture bears witness to the centrality of this reality, and I believe that *God can do what God promises*. But two of the Gospels attempt to explain *how* this incarnation could happen. Dealing with this question of explanation is quite different from description. One cannot even guess how many people through the ages have tripped over the "how," and, in doing so, lost the "what." Matthew's and Luke's narratives are flawed—as explanations. They trace Jesus' genealogy, one from Abraham and the other from Adam, yet do so through Joseph—who they declare was not the physical father of Jesus. Neither Mark nor Paul know of any virgin birth explanation—and yet there is no question about the "what," for they are profoundly committed to the Incarnation of God in Jesus Christ.

So is the case with the doctrine of the Eucharist, over which quarrels have scarred deeply the history of the Church. Augustine's approach was very influential, but misleading in method. As mentioned previously, he insisted that there are an indefinite number of sacraments because the Holy Spirit can act through *anything*. Most Christians would not dispute this. But this dodges the issue. The early Church, in experiencing the Resurrection as profoundly Eucharistic, affirmed that central event to be a sacrament. What this means is that when the Church celebrates this event, God has *promised* to be especially present in communion through the bread and wine. This affirmation in no way limits the action of any of the Trinity, for God can be and do as God wills. *What is at stake is Christ's promise.* We have spoken of the primary sacrament being the Incarnation itself, and derivatively the Church as the ongoing "body of Christ." The Christian is to believe Jesus' clear promise that "where two or three are gathered in my name, I am there among them" (Matt. 18:20). This does not mean that Christ is *only* present there or then, but that he *promises* to be there, and in a special way. *How* this is so is quite another question. Furthermore, if one believes a divine Promise to be so only after having it proven or shown through natural means to be probable, this would not be *faith*. It would simply be a *conclusion*—and people do not die for a conclusion.

My "conversion," as I mentioned in the first book, occurred when the vision expressed with exquisite beauty in the Book of Revelation seized my imagination. The question of "how," which had been my previous stumbling block, was somehow dissolved into a "what if" Jesus had overcome death in the event called "Resurrection." On that important date in my life, the heart of the Book of Revelation came alive, especially for me who was raised in the harsh inequities of Appalachia. Resurrection now meant the vision of a creation that is promised to be—"the World of Sacrament"—like a bride adorned—having the glory of God, its radiance like a most rare jewel, like a jasper, clear as crystal, with walls as pure gold. The streets are to be transparent, and the twelve gates adorned with different jewels of incredible worth. Several things more. He shall wipe away the tears from their eyes, and death shall be no more. And to the thirsty he will give drink from the fountain of the water of life. But this vision *for* the world that *is to be*, has its instrument and foretaste *now*. Since the earth is to be the "World of Sacrament," the Church is now the "soul of Sacrament." That which is, will be. That which is promised, will become.

Could I drink to that vision? *Indeed*—with a chalice. The *how* had become irrelevant. The *what* had become the only thing that would make my life enthusiastically worth living. It rested on one thing: whether Christ's *promise* is to be trusted. Indeed these were Jesus' final words to his disciples in Luke: "And see, I am sending upon you what my Father *promised*" (Luke 24:49, emphasis added).

The next step for me was not long in coming. If the Eucharist is what Christ declared and insisted upon in the Gospel of John, if it is what the early Church claimed it to be in the Letter to the Hebrews, if the Eucharist is the "primal image," the objective meaning and structure of all time and space as witnessed to in the vision of the Book of Revelation, *then* the Church is the Body of Christ where the Eucharistic promise of the past intersects with the Eucharistic vision of the future, as the present in which to participate in Presence. There is no deeper immersion into this Mystery than the Triduum. Here one participates in the Church as the "soul of Sacrament" for the world as sacramental. Fire, bells, singing, immersions, incense, sprinkling, banners, with bread and wine at its core—the experience is that of homecoming. In such an encounter, one is forced toward a *"yes"* or a *"no."* It is as if all of space and time intersect in a moment of destiny on which everything depends. Faith is nothing more, nor is it any less, than gambling on the promise that God is continuing to do in the universe what in the Incarnation God has already done.

The Triduum cannot be separated from the Church—for through it the Body of Christ becomes a holy, incarnate, organic, and living reality. What is more, the Christ who promises his Spirit to the gathered believers as a whole, promises his Real Presence in the sacraments in particular. The Church is the body which "makes" Christ sacramentally Present. In the sacrament, the historical fact of resurrection intersects with the foretaste of the messianic banquet. The Christ who came, and promises to come again, is the carnal Christ here and now. For the eyes of faith, the past becomes contemporaneous through the sacraments as the edge of every future. And these sacra-ments, in turn, issue from the Gospel affirmation: that Christ known originally in physical form continues to be known through the Church as Christ's permanent Incarnation.

Louis Evely notes that all of Christ's relationships with sinners terminate in a banquet with the prodigal son as paradigm. It is in these banquets that we have a "perma-nent epiphany." Every day in the Eucharist, the Word becomes flesh and dwells among us. Since faith is ecstatic and joyful, the Eucharist as thanksgiving flows naturally. And we can affirm in faith that the final consummation of all time and space will be a cosmic Eucharist—for the Eucharist "is so powerful that it is capable of making the whole world participate in the death and resurrection of Christ, making everyone in the world pass from death to life in one definitive Easter."[7]

We can return now to the question with which we began. What are we to do with "transubstantiation," which is the official Catholic position? Aquinas used that term, based on Aristotelian philosophy, in the thirteenth century to help believers *understand* what they *already* believed. It was not intended to be a doctrine one had to be persuaded to believe. To put it another way, "Real Presence" is the *what*, while such terms as *transubstantiation* are efforts to explain *how*. Unfortunately, however, in the conflict of the Reformation, "transubstantiation" became used as a "test question," a catchword for seeing if someone actually believed in the "Real Presence." But even more unfortunate, the *how* was permitted to become the *what*, utterly confusing the situation. What's more, since Aristotelian metaphysics today is largely disregarded, to begin with Aristotelian terminology is almost to guarantee that there will be conflict and misunderstanding. This has become clear in the light of Vatican II, out of which new explanations and new terminology have been attempted, such as "transignification" or "transfinalization." But whatever the theological terminology employed, no explanation of "how" dare be permitted to stand between persons of faith and their lives within the Church as the "soul of Sacrament." In fact, Jesus made clear that "pure faith" is more blessed than any conclusion drawn from evidence (John 20:29). After all, the word *sacrament* does mean "mystery," and so it should remain. Yet the "how" has so often been permitted to displace the "what," that "Real Presence" tends no longer to be believed because of Christ's *promise*. Instead, its acceptance or rejection tends to depend on whether one accepts or rejects an archaic philosophy that is used today only in small circles. However, a shorthand definition remains valid: a sacrament is "a visible sign of an invisible grace." And just as Jesus Christ is the choreographer of Grace, so the Eucharist is the anatomy of the Christ event.

—⊢—

10. The Number of Sacraments

It took a significant part of pre-Reformation history for the Church to settle upon the number of sacraments. The Church agreed early on that a sacrament is that through which God acts. As we had occasion to mention before, since God can act through anything, the conclusion tended to be that there are an indeterminate number of sacraments possible. But that conclusion rests on the premise that the Spirit is totally indeterminate, being *always* like the wind that "blows where it chooses" so that "you do not know where it comes from or where it goes" (John 3:8). Interestingly, in this passage Jesus is speaking of the Spirit's role in Baptism. While affirming the freedom of the Spirit, the Church also maintains that the Spirit is bound by *promise*—to act faithfully in the sacraments. Karl Rahner's approach begins with imaging the Church as the ongoing "Body of Christ." Thus in an originating sense there is one sacrament—the Church, for it is a visible sign of God's mysterious and invisible Presence. Consequently, the Church is the garment of God, and the sacraments are its adornment.

The determination of the number of sacraments proceeded in two contrasting ways. Quite early the Church acknowledged from Scripture two primary sacraments— Eucharist and Baptism—rooted in Jesus' own mandates. Additions to these began to emerge from pastoral practice. After a particular functioning became widespread in use, it might then receive official endorsement by the Church, which would find in Scripture a basis for such endorsement. With the work of Peter Lombard (twelfth century), seven sacraments were basically in place, and the Council of Florence (1439) officially affirmed that there are seven sacraments that "both contain grace and confer it on those that receive it worthily."

It is possible to claim that the Spirit continued to work within the Church to sacralize space-time by affirming a "means of grace" appropriate to each of the hinge-points in the human pilgrimage. For *birth*, then, there is Baptism. For the *coming of age* and *one's self-identity*, there is Confirmation. For *vocation* or *calling*, we have Ordination. For *sexuality*, we have the sacrament of Marriage. For *sickness*, there is Anointing. For *nourishment*, there is the Eucharist. And for *sin*, there is the sacrament of Reconciliation. While from our modern perspective these sacraments as hinge-points might appear too narrow, they were sufficient at the time to constitute the Church as a World of Sacrament. The sacraments involve not only time, but space, in the sense that each uses a physical, tangible element. For Baptism there is water and the oil of the catechumens; in Confirmation there is the oil of chrism; in Ordination, the laying on of hands; in healing, the oil of the sick; in the Eucharist, there is bread and wine; marriage has rings; and in confession, I like to see the tangible element as touch.

In time, however, misuse began occurring—usually instigated by issues of power and greed, as in the case of indulgences. To correct this, as mentioned earlier, the

Protestant Reformers adopted the criterion used by the early Church—Scripture as the index for what could be considered a sacrament. However, since the Catholic Church purported to have found in Scripture a basis for all seven sacraments, this criterion was bound to be less than convincing. While Luther was inclined to include Confession on his short list of sacraments (for sin was central in his theology), in the end he affirmed only Baptism and Eucharist as authentic. The scriptural criterion had two parts: a sacrament had to have its mandate in the words or actions of Jesus, and, in addition, it had to be an action unique to Jesus. Thus, for example, the Reformers did not consider Marriage a sacrament, because it did not meet either criterion. And yet these dual criteria could have rendered "footwashing" a sacrament and removed Baptism because it was used by John the Baptist.

In the pre-Reformation period another divisive issue arose—that of indelibility. For example, if the Eucharistic elements are regarded as permanently changed, certain practices inevitably arose. It was necessary to have a tabernacle for the extra hosts that were consecrated for *viaticum*. Logically possible was the Festival of Corpus Christi, when "Jesus" was placed in a monstrance ("to show") and paraded through the streets. And in the Middle Ages simply adoring the host began to substitute for participating in communion. Thus while Luther and Calvin were inclined to understand the Eucharist in a way very similar to that of the Catholics, they made one exception in an effort to counter abuse. They could affirm that Christ was *in* the elements, *by* them, *under* them, and *through* them—but not *as* them.

In time, however, this difference—which was relatively small in the beginning—became quite large, and in a way came to affect most of the sacraments. The Catholics emphasized the sacraments were "objective," while the Protestants viewed them as "subjective." Put another way, Protestants emphasized what the person brought to the sacrament, while Catholics emphasized what one took away from it because of divine action. Different, too, was the understanding of grace. For Catholics, grace was an enabling power, a strength that came through the sacrament to the recipient. For Luther, however, grace tended to mean God's graciousness in forgiving our sins. As a small Protestant boy, I thought that I would like very much to be excommunicated. Then I wouldn't have to go to church. It wasn't until much later that I understood what a tragedy that would be for the Catholic. It was like never being able to eat again.

The Orthodox Church recognizes the *epiclesis* (praying for the Holy Spirit to come) as the sacred moment when the elements of bread and wine are changed. On the other hand, the Roman Catholic Church before Vatican II recognized the sacred moment to be the words of institution (*anamnesis*). Interestingly, since Vatican II the emphasis is no longer on a particular moment, but on the whole Mass as the instrument of change. This understanding might be more acceptable to Protestantism, for the affirmation is that the Spirit is present in the Eucharist as a whole, and was not brought down at a special place by the power of the priest to do so.

Although most Protestant Churches have only two sacraments, in practice many of the occasions identified in the Catholic Church as being intersected by sacrament have come to be affirmed by Protestants as "ordinances"—meaning "mandates." Deprived by the Reformation of understanding these practices as sacraments, they tend now to be understood functionally. We can illustrate this in terms of *Ordination*. As a Catholic sacrament, an indelible mark is placed upon the person ordained, enabling the person to do what that person could not have done previously. This is so much so that if a priest wishes to "leave" the priesthood, he can be released from priestly activities and deprived of authorization to function as a priest. Yet in no way does that modify the change that occurred at ordination. It is indelible, and thus is for life.

In contrast, most Protestant Churches see no change occurring in the recipient. Instead, ordination has a functional status—it gives the person authorization to function as clergy. Ordination often occurs only after the person has found a parish to serve. If one no longer wishes to continue as a pastor, the certificate of authorization is returned or destroyed, much as one's driving license may be terminated or permitted to lapse. The case of *marriage* is much the same. For Catholics, a marriage is permanent, even if the couple do not understand it that way, and even if married by a non-Catholic clergy person or a secular officer. Marriage indelibly unites the two persons into a new reality.

And if a couple might later choose to separate for reasons of incompatibility, that is acceptable, but it in no way changes their married status. But the Catholic Church has marriage tribunals which, after serious examination, can declare a marriage annulled—as invalid from the beginning. Most Protestant churches agree with Catholics that marriage is a covenant the couple makes with each other before God. But while incompatibility or other causes might induce separation, divorce is also possible in many denominations. This is a bit strange, for if the Church can bestow marriage, the question remains as to why the Church is not involved in its dissolution. What differs among Protestants is the degree to which divorce is considered a "sin," for which it is appropriate to ask forgiveness from God. Some churches regard divorce as regrettable, and yet can affirm it as a positive step in the pilgrimage of both individuals.

There is a significant difference presently taking place within Protestantism with regard to *Baptism*. Increasingly it seems to be difficult for mainline Protestant churches to affirm that in infant baptism an actual, objective change is effected. A few churches have taken the consistent next step of acknowledging this act as a "christening" or "dedication" of the child. For other churches that retain infant baptism, the tendency is to move toward the subjective pole, seeing the act as a recognition of the sacredness of birth, an acceptance of the child into the Church, and a promise by parents and the Church to do all in their power to "increase [the child's]

faith, confirm their hope, and perfect them in love." [8]
Baptists, on the other hand, insist upon believer's baptism,
which is an act of promise on the part of the "born again"
Christian. Baptism is not regarded as a vehicle for this
change, but the Church's acknowledgment of the change
after it occurs. Thus the number of sacraments depends
upon the denomination. They range from the Reorganized
Church of Jesus Christ of Latter-Day Saints, who have
eight, to the Quakers, who have none. Yet the Protestant
churches, while affirming two "official" sacraments, are
moving toward providing explicit ministry at the hinge-
points of life, thus opening the opportunity for sharing
with the Catholic tradition.

—⊢—

11. Liturgy as the Context of Sacrament

In dealing with "eternal time" in the previous book, we
indicated a general structure of Protestant worship. In
many Protestant churches, the sacraments are short liturgies
inserted into the general worship service as needed. The
Roman Catholic Church functions similarly in the sense
that it inserts into the general worship structure of the
Eucharist other sacraments as needed—making for a sacrament within a sacrament. The revised liturgy of the Mass
since Vatican II has two primary parts, enclosed within a
short introductory rite of forgiveness and a concluding rite
of dismissal into the world. The first part is the "Liturgy of

the Synagogue" or the "Liturgy of the Word." Most early Christians were Jewish, and therefore continued to worship at the local synagogue. When they were no longer welcome there, they incorporated into the Christian context this Jewish type of worship, which was composed of Scripture, exposition, and prayer. The second part of the Catholic service is the "Liturgy of the Upper Room," or the "Liturgy of the Eucharist." In this part, Vatican II resurrected some of the ancient liturgies, one of which is an Eucharistic prayer dating back at least to Hippolytus (fourth century). The priest can choose from three other primary Eucharistic prayers, with further options for special feasts and events.

The more liturgical a church, the more does the intersection of space and time find affinity with the artistic "world." Both celebrate life through color, shape, and rhythm. Vatican II has bequeathed a fine format for the Mass as a liturgical drama. What is needed now is to bring increasing imagination and feeling to that form. Protestants, on the other hand, need to become more theologically aware of the liturgical structure operating behind the loosely held ingredients of the worship service, and determine the degree to which their "drama of the Word" might intertwine with the drama of the Eucharist. In so doing, we can hope that these denominations might regain and contribute some of the important meanings on which they were founded.

As we identified previously, Catholicism and Protestantism are contemporary versions of two broad liturgical traditions discernible throughout the sweep of Scripture and tradition. One tradition is that of the *prophetic Word*, linking together Abraham, the prophets, Paul, and the Protestant Church. The second tradition is that of the *priestly act*, linking together Moses, the Levitical priests, Peter, and the Roman Catholic Church. Unfortunately, these necessary traditions tend to be posed in either/or fashion. In fact, my training in a Protestant seminary identified the prophets as the faithful ones, while the priestly tradition was largely neglected, discharged as unimaginative and backward keepers-of-the-status-quo. It is becoming increasingly clear that both traditions are needed and that when they are separated, the Church appears to be incomplete. Thomas Merton saw the importance of these two great churches bringing together the Protestant concern for the sanctification of one's soul and the Catholic representation of the whole Church before the throne of God. For the Protestant, Holy Communion partakes of the intimate space of being invited to the family table; for the Catholic, the space at the altar in the Eucharist participates in an infinity of Mystery. We need each other.

This accounts in large part for why Vatican II has been so important to Catholicism. On the one hand, it recovered the purity of the priestly tradition, using as the norm the liturgy of the early Church. On the other hand, the prophetic tradition became mandated, requiring a proclamation of the Word at every Sunday celebration of the priestly act. As an indication of such change, there is currently a tendency in Catholic architecture to make the lectern for Scripture symbolically of a similar height as the altar. On the other hand, in the mainline Protestant churches one can detect a hungering for a renewed priestly emphasis, centering on increasing the place and frequency of the Eucharist. One Catholic encouragement of this renewed interest among Protestants is this: "When the Church becomes a sacramental community, no matter how bad the sermon, we still get to break bread with the Lord."

This interaction between the two great traditions is to be welcomed. In *My Fair Lady*, Catholics might well identify with the heroine who in frustration shouts, "Don't talk to me of love, show me!" And the Protestant response might well be, "Don't just show me; speak to me of love!" In the Liturgy of the Word and the Liturgy of the Eucharist rest the powerful diversity and richness of Christian worship, for they are meant to intersect with one another. I have experienced them in the contrast between the Trappist Liturgy of the Hours and the liturgy of the "High Mass" at a Cathedral. At the Liturgy of the Hours, the mood is quiet, pondering, reflective—the congregation chants the psalms as the Church has done for nearly two thousand years, climaxing in the simplicity of the Word read. No adornment is necessary, for the senses bow to the sense of hearing. This is also the aura of a Quaker meeting, where the hearing is inward, or the gentle simplicity of the furnishings of the Shakers, or the Appalachian clapboard gospel church on the hill, by the cemetery. One feels powerfully the lonely soul disciplined in faithfulness.

In decided contrast, yet equally important, is the liturgical worship of the "High Mass," in which artists are invited to give creative form to the chaotic as the liturgy participates in the struggle of Being with Nonbeing. The interplay of rite and ceremonial and rubric and order emerges—and mingles with matter, form, intent, and disposition. A choreography of physical expressions emanate—of kneeling, standing, and sitting, with the gesturing of embracing, reaching, enfolding, and praying. The beauty of dance is hinted within the presenting, blessing, breaking, and offering—of Christ poured forth. Around the altar flow graphic colors and rhythmic music, the movement of vestments, stoles, tapestries, banners, altar and lectern appointments. Even a flamboyance of singing and instrumentation might be risked—from flute to drum, separating the sounds with a delicious silence. Then might the Word be shared, punctuated imaginatively with a pillar of fire, a cloud of incense, imposed ashes, with visual projections of when and where. Light and dark flow into moving shadows, as the smell of smoke intermingles with the fragrance of incense and flowers. Welcome is the touch of objects, and of persons—of wood and ceramic, of metal and glass—sculpted and carved and stroked—a hand held, an embrace. Words might be shouted and whispered—across and up, down and over. The Sacramentary and Lectionary, the two liturgical books, could be scribed and bound by loving hands. And water there will be—in abundance, blessed and dipped, touched and sprinkled, held in the flowing lines of font and cruet.

About a month ago, I celebrated Eucharist at the altar of our new sanctuary. As I held the chalice for the blessing, I looked down into it. Shining gold bottom, pink wine, ceiling lights sparkling and reflecting against the sides—I was mesmerized by the beauty—with the blood of Christ fit for heaven and earth. Kathleen Norris wisely observed that monastic persons seek to weave ceremony through every mundane part of life—how one eats, dresses, treats tools, and enters a church.[9]

The mature Merton came to understand the sacramental liturgy as a state of being and a way of living. When he rewrote his classic, *New Seeds of Contemplation*, he added a short chapter called "The General Dance." The world is a garden in which God takes delight, he said. We are the artists, workers, and gardeners of that paradise, so loved by God that it is impossible for anyone really to be God's enemy. In the past, we acted so as to render God a pilgrim and exile in his own creation. But there can come a pure moment in which we can see clearly that what God truly wants is to play with us. We can catch a clue to the game "when we are alone on a starlit night," or see "the migrating birds in autumn," or when children are children. Everything can "provide a glimpse of the Cosmic Dance" that "beats in our very blood."[10]

Liturgy is play. Life is liturgy. So it is that the "cosmic dance" of the universe and the Eucharistic gestures of the Church's liturgy know that they have been made for each other. And as play, perhaps the game may be Hide-and-Go-Seek—with the divine call being "Olley, Olley, in come free."

4

SACRAMENTS AND SACRAMENTALS

"How could we sing the LORD'S song in a foreign land?"
(Ps. 137:4)

"I have come into deep waters." (Ps. 69:2)

"The LORD lifts up those who are bowed down; the
LORD ... upholds orphan and the widow." (Ps. 146:8-9)

1. Baptism: Catholic and Protestant

From the inception of Christianity, the faithful have practiced the sacred event called Baptism. Jesus' final words
called his disciples to make "disciples of all nations," by
"baptizing them" and "teaching them" (Matt. 28:18-20).
Yet several issues have arisen that Christians have debated
ever since. For example, should Baptism be done first, to
be followed by teaching, as would be the case with "infant
Baptism"? Or is teaching first, with Baptism a witness to
the "state of grace" in which one comes to stand—as tends
to be the case in "believer's Baptism"?

A basic question arose from the Protestant criterion
for determining if an act is a sacrament—namely, the need
for it to be "unique" to Jesus. Baptism was not unique, for
it was central to the ministry of John the Baptist as well.

Partly as a result, tradition distinguishes between John's form of baptism and that characterized by the life and death and resurrection of Jesus. In fact, this difference continues in emphasis, distinguishing the view of Catholics and Protestants. Baptism for Protestants is close to that of John, for both men preached, "Repent, for the kingdom of heaven has come near" (Matt. 3:2). For reasons largely based on their negative view of the human condition, the Reformers stressed Baptism as a washing, a forgiveness of sins. But when compared with John's approach, they also stressed Baptism as a work of God's graciousness, rather than "works"—as an act we initiate by repenting. Hope rests in God's forgiveness of sin as a free gift of Jesus, who died for our sins, as appropriated in Baptism. One needs to be washed clean—purged for a new beginning. When Baptism of believers is stressed, Baptism functions as a sign of something *already* so, as a human witness to a person already justified by faith. When infant Baptism is practiced, there has been difficulty in Protestantism knowing how to speak of the forgiveness of a child. This difficulty, in part, has led many Protestant Churches today to treat Baptism as a sacramental. As a result, the difference between Protestant and Catholic again illustrates the traditional distinction of a sacrament as *causative*, and a sacramental as *evocative*.

While Catholics are affirming of the Protestant emphasis on Baptism as *forgiveness of sin*, their distinguishing emphasis is increasingly upon Baptism as a *sharing in Christ's death and resurrection*, removing the threat and fear of *death*. While infant Baptism had been the most used form, since Vatican II the Roman Catholic Church has attempted to bring the physical act of Baptism into conformity with its meaning. From early times, the baptismal act was *immersion*, in which the candidate, standing in water, was taken backwards into the water (and thus out of control) three times—as Jesus was in the tomb three days. Each time the person was "at the mercy" of the priest, trusting that he would "raise up" the person. When finally raised up as if on the third "day," one is given a baptismal name, and the senses are sealed with holy oil against temptation. Consequently we are baptized *into Christ, with him in the womb of death, rising with him from the tomb*—with death behind, and the transformation of time still ahead as a pilgrimage into eternal life. As we are raised, we gasp, breathing the Holy Spirit into our aching lungs.

The Triduum, as the celebration of Christ's death and resurrection, is the most appropriate time for Baptisms, especially of adults, because they celebrate Christ's death and resurrection. Even if a particular Catholic Church does not have anyone to baptize during the Triduum, a central part of the event still deals with Baptism in the renewal of baptismal vows. The Paschal Candle, while important throughout, has an interesting role in making holy the baptismal water. Three times it is plunged into the water in what may be the most explicitly sexual act in Christian liturgy. This renders sacred the act of conception—as it was with the Holy Spirit, Mary, and the Incarnate One. It symbolizes rebirth from the seed of the

Spirit, and the womb of the font. Then comes an exu-
berance of dipping and sprinkling of water, everywhere on
everyone. Baptism, being indelible, is not to be repeated,
but functions as the divine promise that all of life is an
ongoing Baptism—of life over death. This suggests its link
with Confirmation, which marks a new beginning in one's
unique lifelong vocation. Thus three of the sacraments—
Baptism, Confirmation, and Eucharist—are regarded as
the sacraments of initiation, functioning as the first
installments of our promised inheritance.

An important part of the Vatican II revision was the
development of a prebaptismal preparation program called
RCIA ("Rite of Christian Initiation of Adults") for the
Baptism and Confirmation of adults. It usually takes at
least six months of weekly sessions, with each candidate
teamed with a member of the congregation. This thorough
educational program begins with those interested in
learning more about Catholic Christianity and provides
several built-in places where a person might opt out of the
program. At each point there is a liturgical observance as
persons move closer to commitment. Holy Week is very
intense, with the crescendo (following the practice of the
early Church) being at the Easter Vigil. The nine days after
Easter are special days in which new members are
instructed in depth about the sacraments—for they are
now a living part of the Church as the World of Sacrament.

The contrast between Baptism according to John and
that of the early Church does not emerge clearly in the
Gospels. In fact, at the beginning both John and Jesus

seem to have the same mission. "Jesus began to proclaim, 'Repent, for the kingdom of heaven has come near'" (Matt. 4:17). While the Gospels are structured so as to be read from beginning to end, the perspective from which everything was written was "end to beginning." This accounts for why the disciples are portrayed as not being very smart—they "don't quite get it." That is because *we* know how it will all work out, and what the end is that alone gives meaning to the whole. They did not. John baptized Jesus at the beginning of his ministry, but we are not baptized as Jesus was. Our Baptism is "into Christ." It is the Triduum that flows back over the whole of Jesus' life—providing the meaning of the two primary sacraments, Baptism and Eucharist.

In the monastic setting the office of Compline ("complete") signals the end of our life each evening, for life is given in one-day increments. At that office, it is as though we repeat the baptismal act of dying with Christ. If the next morning we are re-given the gift of resurrection, we rise to new life, completing the remembrance of our Baptism. It is hard to give up our lives, and in order that we may do so with courage, at the end of Compline we are sprinkled with water as a sacramental, to remind us that in Baptism our death has already occurred. It reminds us that our life now is a foretaste of the resurrection that is the ultimate promise of Christ for all reality. The new name given at Baptism bears witness to a whole new state of being.

Reflecting the triad of events through which the Christian understands all of time, the triad of the new life of Baptism is that of dying with Christ, being united in burial with Christ, and rising with Christ. In a real sense, Baptism is a reentry into the womb—not the biological one this time, but a new one, the womb of "Holy Mother Church." This event stands midway, so to speak, between Eden and the Kingdom, between Advent and Epiphany, between Lent and Pentecost. This whole understanding is appropriated in monasticism. When the monk makes lifelong vows, he enters a second Baptism. Prostrate upon the floor, he is covered with a black cowl of death. Then he is raised up, to a newness of life, and given a new name, the name of a saint who is to be one's lifelong model.

One can see here how the Catholic funeral is an extension of one's Baptism. A Christian is baptized in a pure white gown (actually or symbolically), which one keeps. As one's casket is brought into the entrance of the Church, it is covered with that very gown or with a pall (a white cloth) representing that garment. The casket is totally covered, which also negates the difference between a gold-trimmed casket and a pine box. Its being covered signifies that one has already died in Christ, and in that promise one's Resurrection is fact. This understanding, new after Vatican II, brought a significant change in funerals. No longer is this event identified as a "Requiem Mass" or a "Mass of the Dead," with black the color used for vestments and paraments. Instead, it is now the "Mass of the Resurrection," with the color being white—the color of Christmas and Easter. The huge Easter Candle which was first lit at Easter Vigil is rekindled at funerals and leads the procession of casket, servers, lector, and minister into the church as an Easter hymn is sung, such as "Christ the Lord Is Risen Today." The casket is placed feet first toward the rising sun, from which direction Jesus' second coming will occur.

Before Vatican II, infant Baptisms were often private. Now the very opposite is true. Baptisms occur during a major celebration of the Eucharist, in which there is an extensive calling of the saints by name as they are invited to be present as witnesses. The revised liturgy picks up the theme of water as it has appeared throughout biblical history—the Spirit moving over the waters at the dawn of creation, the great Flood, liberation through the Red Sea, Jesus' own baptism, and the water and blood flowing from his side. Protestants, on the other hand, tend to interpret this initiating event with images of washing and cleansing. Judged in one's self-centered state of sin, one is forgiven and then set free. In a blue-collar Protestant church I once pastored, together we found analogous meanings in how we actually use water daily—pouring as in the watering of plants for growth, sprinkling as with laundry so as to iron out the wrinkles, and cleansing as in washing one's hands before a meal.

The Scriptures are full of baptismal imagery. "I have called you by name, you are mine" (Isa. 43:1). Baptism is an engrafting and a pruning. It is an act of uniting with Christ, and walking henceforth with him. It is a planting of the seeds of the Holy Spirit, "daily" watering with holy water. It suggests the water of one's primal birth, in this case, a second birth. Baptism is the incarnation of Christ within one's soul, and thus is a force throughout one's life. After Baptism, the early church gave the person a drink of milk and honey, as foretaste of the Promised Land. Baptism is continuous with God's covenant with Israel, expanded now into the universal Church, with each Christian becoming part of this Promise. Out tumbles more baptismal poetry as act: to be marked, engrafted, cleansed, claimed, graced, sealed, called, plunged, covenanted, embraced, dedicated, rendered unbreakable, reborn, made new, to become part of the body of Christ, to begin again, to die and rise, as a result of the divine initiative. Then flows the human response—by the person, or parents, and later promised again in Confirmation—to confess, intend, begin again, commit, obey, and dedicate.

Confirmation is a sacrament in need of a theology. It can be characterized as one's personal Pentecost. And yet there is a theological problem. Baptism is understood not only as a baptism with water, but also by the Holy Spirit. Therefore I find it more helpful to speak of Baptism-Confirmation, hyphenated as two parts of a whole, usually separated by time. In that sense, Confirmation preserves the priority of grace, while making room for a willed response—both necessary for the fullness of the sacrament. This could bring together the two views of Baptism—the priority of grace as with a newborn child, and the importance in "believer's Baptism" of a person's response to forgiveness. Confirmation is not a gift *from* the Holy Spirit, but is a gift *of* the Spirit. If Confirmation is interpreted as an act of maturity, this would give time for the fuller dynamic of incorporation into Christ—to relive his life, death, and resurrection, as symbolically done at Baptism during the Triduum. Just as we have insisted on linking *promise* to each sacrament, so Luther insisted on Baptism as the divine Promise that is in force all of one's life. Thus, for him, sin and forgiveness are to be understood and lived out through faithfulness to one's baptismal commitment.

As is the case with every sacrament, Baptism involves the Word and a natural element (water), intersecting as divine Promise. Actually, then, both "dramas," the Catholic and the Protestant, point to important aspects of Baptism—death and resurrection, and sin and forgiveness. Both are needed if we are to catch the fuller meaning of Baptism. Combining images, Baptism is a symbolic immersion, a washing and bathing, a grafting into the Paschal mystery, an adoption into the Body of Christ, total forgiveness as entrance to a new life, initiation as entering into community—an indelible oneness with the life, death, and resurrection of Jesus Christ, as the beginning of a lifelong vocation.

—⊢—

2. Baptism and Ecumenism

In the summer of 1999 at Ghost Ranch, New Mexico, a prominent Presbyterian lay leader and I conducted an ecumenical encounter on the two major sacraments, Baptism and Eucharist. The intent was to determine in practice how much agreement there might be among the "denominational" families of the Church and how much dissension would emerge. Those in attendance represented a considerable spectrum, from Unitarian Universalist to Roman Catholic. After intense engagement, as well as periods of backing off, we finally were able to construct a statement on Baptism that, surprisingly, all in attendance could agree upon. This was not necessarily a complete statement, but at least the group found it acceptable as far as it went. When this was completed, a second effort attempted to identify the additional elements that one or more of the denominations would need for emphasis or completeness but about which there was not common agreement. What follows is the "Common Baptismal Statement."

> Responding in faith to Jesus' instructions, the Church as the body of Christ baptizes people in the name of the Father, Son, and Holy Spirit. In this sacrament, God acts through the rich symbolism of water to engraft people into Christ. They participate in his death and resurrection and, having now died and risen with Christ, they are claimed, marked, and sealed as belonging to God forever. This act of grace

initiates a journey of discipleship in which those baptized and cleansed from sin follow Christ in obedience to his call to be part of a new covenant relationship with God. The congregation in which this permanent and unbreakable act of Baptism occurs dedicates itself to nurture and sustain those baptized, including children, whom it helps to understand that they belong to this community of faith and one day may confess their own faith.

Although different Christian traditions use different forms of Baptism, in each case the church understands the person baptized to be plunged into the embracing love and family of God. Also, Baptism is the way by which people say a decisive yes to the gospel and by which God says a decisive yes to those baptized—and then sends them out to live and die for the transformation of the world. Key words identified by the group as possessing important symbolic value for understanding Baptism were these: engrafted, die and rise, adopted, cleansing, indelible marking, calling, abiding in God forever, unbreakable, marked as God's own, belonging, and plunged.

It turned out that additions needed beyond the common statement reflected not issues between Protestant denominations, but additions needed from a Catholic perspective—with the Protestants present being satisfied with the common statement. The following are aspects from a Catholic perspective, identified either as needing emphasis or as additions:

1. The need for a clear *distinction* between Baptism as a "dedication" and as a "sacrament";
2. The organic nature of the person's *incorporation* into Holy Mother Church, as crucial for affirming the organic nature of the Body of Christ;
3. Making a significant *relationship* between Baptism and Confirmation, emphasizing the ongoing role of the Holy Spirit in one's pilgrimage as a growth in grace;
4. Baptism as the *cause* effecting divine forgiveness and incorporation;
5. The primary *imagery* for Baptism as that of Christ's life, death, and resurrection, yet forgiveness as an important aspect;
6. Baptism as being *branded* by the Holy Spirit, owned by God, and never to be undone.

—┬—

3. The Eucharist: A Search for Understanding

As noted earlier, the Eucharist is the central sacrament, but it has become the most complex because of heavy disagreements between denominations, rendering any agreements ambiguous. Denominations do not even agree on the name. It has been variously named the Lord's Supper, the Breaking of Bread, the Divine Liturgy, Holy Communion, the Lord's Memorial, Mass of the Faithful, and Service of the Table. Since the first century, however, the term *Eucharist* has been used, and with its meaning of "giving thanks," it carries less baggage than other names.

As the Eucharist has been celebrated through the centuries, eight themes can be identified:

1. Thanksgiving
2. Communion (fellowship together and with Christ)
3. *Anamnesis* (a re-remembering)
4. *Epiclesis* (invoking the Holy Spirit)
5. Eschatology (anticipation of the new heaven and the new earth)
6. "Medicine of Immortality" (granting eternal life)
7. Real Presence
8. Sacrifice

It is around these last two dimensions, "Real Presence" and "Sacrifice," that disagreement has been most vigorous. What different sides agree upon is that Christ's sacrifice on the cross is both central and "once-and-for-all." They might not argue over Christ being both "priest" and "victim," as suggested in Hebrews. They might find some commonality on the issue of the costliness of forgiveness. They probably would not disagree much in identifying the Eucharist as an act of eating and drinking, in thankful remembrance and joyful anticipation. We could jointly affirm the practice of the early Church in which the elements for Eucharist came from homes, by participants dressed in the ordinary clothes of everyday life. We could possibly agree that while the Jewish tradition tended to see the ordinary as unclean, the

Eucharist makes the ordinary sacred—transforming ordinary things, ordinary relationships, and ordinary places. To make this aspect a living truth for most of us, Christians might well celebrate Eucharist in living rooms, in places of vocation, on the workbench, or even in the sandbox.

The United Methodists have modified their abolition of alcoholic beverages, making possible the use of wine in the Eucharist. To share in the joy of living with the psalmist, we too might together make glad and free the human heart. We might jointly recognize as valid a social justice theme, discerning the Eucharist as a dress rehearsal for the first supper in the new age. Perhaps Protestants could be sympathetic regarding terminology by understanding that the Catholic term *Mass* comes from the word *dismissal*, indicating that the purpose of the Eucharist is to send believers out into the world as leaven. Could we acknowledge the deeply corporate nature of the Eucharist, recognizing it as a catharsis for the terrifying loneliness in our time? Then we become restive, but could we push a bit further in together recognizing the confluence in the upper room of "when," "what," "why," "for." Here we reach the place where two additional steps still might be possible. Could we join together in participation within the Triduum, from which all things flow? And again, could we agree on the "promissorial" nature of the sacraments, as we have been developing throughout—namely, to trust the promises of Christ, even though we might never agree on the "how"?

—┬—

4. Adaptability of Eucharist Liturgy

Churches are most likely to be right in what they affirm, and suspect in what they deny. Before Vatican II, the Mass was seen primarily as a "sacrificial" event, offering up the body and blood of Jesus to God for the sake of atonement. The altar was against the far wall, with the priest speaking low in Latin, his back to the congregation. Since Vatican II, however, this emphasis has been minimized. The heavy stress now is on "communion." For "communion" to be acted out, the priest must be able to stand behind the altar, thereby making it possible for the "altar" to function as a table. The priest faces the congregation at all times, and when the congregation is small, all can stand around the table in fellowship, now participating in English rather than Latin.

Interestingly, Protestants have always done what Catholics are now doing, using the language of "communion" as the nature of the whole Eucharist, rather than just the time of eating and drinking. The Protestant name is usually "table," and not "altar." In fact, so much has this sense of fellowship been central for Protestants that in some churches the communion table is extended as rails, enabling twelve people to kneel and receive at one time.

Having been down the road that Catholics are presently inclined to follow, some Protestants are reacting against the insufficiency they feel, by moving in a reverse way. That for which such Protestants seem to be seeking is an "objective" dimension, in which Holy Communion is a

sacred *event*, and thus more than fellowship. The key difference formerly resided in whether the primary underlying
doctrine is incarnation or atonement. Lest we just reverse
"end zones," we need each other. What if the latter is for the
sake of the former? Eucharist, then, becomes the sacrament
of resurrection.

In conversation with my Roman Catholic bishop, I
expressed this concern over what we may be losing in the
attempt to regain an important dimension in understanding the Eucahrist. Mustering my courage, I asked what he
would say if I tried to honor both traditions. My idea was
that, after the Introductory Rites, Part One of the Mass
(Liturgy of the Word) would be done as it is presently,
which is akin to the synagogue worship from which it was
drawn. My suggestion had to do with Part Two, the Liturgy
of the Upper Room. The change would be for the priest to
begin this event by facing the altar from the front, standing
with the people. The old form was interpreted by lay persons
as the priest "having his back to the people." My idea is to
discover a way to convey not that the priest is ignoring the
people, but is standing *with* them. A climax would come
when the priest, still facing the altar, lifts high the gifts to
God, with the words: "Through Him, with Him, in Him,
in the unity of the Holy Spirit, all glory and honor is Yours,
almighty Father, for ever and ever." Everyone would sing a
thunderous "Amen." We have participated in the sacrifice.
Then the priest would walk to the other side of the altar-
table, this time facing the people, and say: "Having made
our sacrifice to God upon this altar, it becomes a table

around which we gather now in communion, praying the table blessing that Jesus taught us." All of us would pray the Lord's Prayer, and continue on to the end, as presently done.

The bishop thought over my idea for a moment, and replied: "I think that captures very well the meaning of the Mass. Why don't you try it?" I did. It was well received. Many said, "I understand the Mass better now!" But interestingly, the hardest part for some was "going back to having the priest turn his back to us." It doesn't take long for changes that we thought would be difficult to become normative. Now the question that emerges is whether it would be possible for Protestants, coming from the opposite direction, to meet with Catholics near the center? Protestants have almost always had the minister standing behind the "Communion Table." Would it be possible now for the opening part of the Protestant Communion Service to be done with the presider facing the "altar"? Perhaps not. All this is to clarify the degree to which formation and instruction are crucial to make clear the intention at stake in the changes. In order to symbolize this concept of the priest with the people, some congregations are experimenting with placing the altar in the midst of the people. In a related move, some churches have the prayers of the people raised from within the congregation itself.

Catholics are more restrictive in the use of the word *communion* than Protestants. It refers only to the actual receiving of the elements. Presently, non-Catholics are invited to participate in every part of the Mass except "communion." Actually everyone is invited to come forward, with the non-Catholic invited to cross his or her arms over one's chest and receive a "blessing" from the priest—infants and children included. Another difference is that in many Protestant churches the elements are served in a manner that Catholics call "cafeteria" fashion—as self-service. In contrast, the Catholic is "given" communion by the priest. He holds up the host for the communicant to see, with the words, "The Body of Christ." For him it is an affirmation, to be received by the communicant as a question: "The Body of Christ?" The person is to respond "Amen," meaning "Yes!" Only then is the host given.

The extent to which Catholics and Protestants are coming together in regard to the Eucharist might be measured by how much ecumenical agreement exists to Vatican II's statement in the "Constitution on the Sacred Liturgy, No. 47."

> At the Last Supper, on the night He was betrayed, our Savior instituted the Eucharistic Sacrifice of His Body and Blood. He did this in order to perpetuate the sacrifice of the Cross throughout the centuries until He should come again. For this He entrusted to His beloved spouse, the Church, a memorial of His death and resurrection: a sacrament of love, a sign of unity, a bond of charity, a paschal banquet in which Christ is consumed, the mind is filled with grace, and a pledge of future glory is given to us.

For the Catholic, Eucharist serves as an expression of a unity *already* attained. Relatedly, some Protestants practice "closed communion," requiring membership of those who have undergone "adult Baptism" in the denomination. Quite interesting is the stand of United Methodists. Wesley understood Eucharist to be an evangelical or converting sacrament. Since through Eucharist one is empowered by the Holy Spirit, to receive communion is a crucial way in which the Spirit may bring about a conversion to Christian life. But because of the difference of churches on this issue, the struggle over the *meaning* of Eucharist will have to be done without Eucharist as a means toward unity. But may the Spirit birth for all of us the humility no longer to accuse Protestants of only having a "Kool-Aid party," or of accusing Catholics of practicing "magic."

—⊢—

5. Eucharist: An Ecumenical Exercise

After the denominational participants at Ghost Ranch had completed their work on Baptism, an even longer time was needed for dealing with the Eucharist. A moving experience was the willingness of a Catholic priest to explain the Catholic stand on intercommunion, then celebrating the Eucharist, making clear the meaning of the parts as he presided. The statement that the denominational representatives at Ghost Ranch were able to create and agree upon was this:

Jesus established the Eucharist as a means by which he could be present, after his death, to his followers in every time and place. These followers experience the benefits of his promised presence at his table by being in communion with him and with one another, even though they may feel unworthy of his grace. Christians celebrate this meal in many ways and with different understandings of its meaning, but in some mysterious way, Christ draws all variety into himself as an answer to his prayer in Gethsemane, that all his followers may be one—sent to live out his love in the world.

The following are the "additions" that the Catholic perspective wanted to add or to underscore.

1. The Eucharist *effects* what it symbolizes.
2. The *distinction* between sacrament and sacramental needs to be made and kept clear.
3. The Eucharist is based upon the *promise* of Christ and the Presence of the Holy Spirit.
4. Christ's Incarnation, Crucifixion, Resurrection, and Ascension are once and for all, but in such a way that they are *ongoing* in time and space.
5. Eucharist entails *participation* in the divine.
6. Eucharist is the heartbeat of the *Church* understood organically as the Body of Christ.
7. The resurrection occurs *ongoingly* when Christ is known in the breaking of the bread.

8. The key to the Eucharist rests in the *"what"* rather than the "how."
9. The *Presider* at the Eucharist must be ordained, serving *in persona Christi* ("acting in the person of Christ").
10. *Tabernacle* is the place where hosts are retained in a designated place to be available for *viaticum* (food for the journey of the dying).

I received affirmation of one statement more: that these elements or dimensions are implicit within the event of *Triduum* as the primal meaning of time and space.[1]

—⊢—

6. Post–Vatican II Possibilities

As a way of seeking understanding between Catholics and Protestants in regard to the Eucharist being the center of both Church and world, it is helpful to consider some of the factors and situations that have occurred since Vatican II, over thirty years ago, that may indicate significant movement toward mutual understanding.

Radical Change. The Catholic Church, after initial confusion and a variety of emotional responses to Vatican II, has shown that radical change *can* happen, by which one can recapture a purity of tradition. To give an illustration, in the medieval period worshipers were encouraged to come forward for communion at least once a year.

Presently, as directed by Vatican II, almost all worshippers come forward every week. A similar witness is the vast change in the Episcopal Church over the past thirty years. While Morning Prayer used to be the primary Sunday worship, it is now the Eucharist—with Morning Prayer viewed as a lesser alternative when no priest is available.

The Moment of Change. The Eastern Orthodox approach stresses that the real change occurs in the Eucharist at the *epiclesis*—with the invoking of the Holy Spirit ("Send down Your Holy Spirit..."). The Catholic approach before Vatican II stressed that the "change" in the elements occurred at the Words of Institution ("On the night he was betrayed..."). Since Vatican II, the Catholic Church understands that the "change" does not occur at any one specific time. Rather, the Eucharist, taken as a whole, is when, where, and through which the Holy Spirit instills the meaning in which we participate. Since Protestants tend to have a high doctrine of the Holy Spirit, and recognize the Words of Institution as what makes Holy Communion more than a sparse meal, we might be within hailing distance of one another.

The Holy Spirit as Enabler. The mature Calvin did not argue about the change in the elements, insisting, rather, that because the Holy Spirit works through them, they are efficacious. Put in terms of his own Eucharistic liturgy, we are "partakers of his body and blood in order that we may possess him wholly and in such wise that we may live in him and he in us."[2] Eucharist is participation in the divine action of the Spirit. When pushed, Calvin spoke about Presence in spatial terms—that Christ is present "in, with, and under, the bread and wine." The only preposition missing was "as." Most scholars now recognize that omission to be minor, understandable in terms of Calvin's intent on faithful reform. Another simple sentence from his Eucharistic Liturgy is promising for ecumenical agreement: "We feed on Christ through the mystery of the Holy Spirit."

Divine Presence. Vatican II encouraged a broadening of the ways in which the "divine Presence" is to be recognized—such as in the person of the priest, in the bread and wine, in the whole sacramental action, in the congregation, and the world itself as divine sacrament.

Eucharist and Response. Protestants are strong on "human response," and often uneasy about divine action in which the recipient appears to be passive. But with Vatican II there has been an understanding that we offer ourselves with Christ in his offering. This is the meaning of coming forward to receive—as an offering of oneself.

Transubstantiation. This term is a red flag within ecumenism. In part this is because the Reformers focused their harsh judgment on this term. For example, in Article XVIII of the United Methodist "Articles of Religion," not only is there a rejection of the term *transubstantiation*, but it is regarded as "repugnant to the plain words of Scripture" and "hath given occasion to many superstitions." In Article XX, "the sacrifice of mass" is rejected as "a blasphemous fable and dangerous deceit."[3] There is little to be gained by returning to the anti-Catholic sentiment that was lively during the period of the Reformation. While this term still appears in Catholic documents, Vatican II encouraged alternative terms and ways of expressing the meaning to which the dated word *transubstantiation* was meant to point. Such terms as *transfinalization* and *transignification* are among others that are currently in use. The latter term, for example, emphasizes the "change" as being not so much in the "material" of the Eucharist as it is in the "meaning." By equating "being" with "meaning," we have a very different approach, and one that should be understandable to Protestants. This approach can connect directly with the work of the famed early-twentieth-century Protestant theologian, Albrecht Ritschl. Knowledge of things religious is determined as true, he affirmed, in regard to their value or "worth for us." Thus instead of insisting upon narrow terminology made up of several test words, what is needed is to flood the Eucharist as the *Event of events* with poetry, imagery, simile, metaphor—finding in that rich collage a way of sharing the divine-human meaning.

Meaningful Gestures. Protestants often criticize what appear to them as "meaningless gestures" that the priest is "required" to do, which end up distracting from the meaning of the Eucharist. Historical research into origins and meanings, undertaken in conjunction with Vatican II, has helped indicate which gestures, words, and meanings are duplicative and need to be eliminated, which have lost their meaning and need exclusion, and which need to be kept or restored. The result is a liturgy rich and fresh with symbolism.

Signification. A key norm for the Catholic is that the Eucharist, in being a sacrament, is a symbol that *produces* what it *signifies*. On the basis of what we have already indicated, this issue might be capable of being resolved, but as of now, it stands as a significant disagreement.

Communion as Social Justice. Protestants often pride themselves for their commitment to social change. In the last decade or so, third world theologians have helped us see that the Eucharist entails "the unconditional sharing of bread." for all, especially the hungry. Thus there should be ecumenical appeal to understand the Church as calling for a classless society, in which the Eucharist is the beginning of a continued economic and political movement throughout the world. In fact, simply to receive Communion is to be in identification with the poor and marginalized. The Greek word *anawim*, meaning "God-beloved poor," is also the word for "grapes," crushed for the Eucharist.

Applicability. To be centered in *the* Eucharist is to be able to dance everywhere, in the vast incarnation of Real Presence. Today we need, above all, to be grasped by the sacrificial dimension of the Eucharist. Crucifixion is an utterly beastly business. It rivets our souls to picture the race to which Jesus belonged, standing helplessly in line at Auschwitz to enter the death chamber. The sound of the fracture of the bread is like bones broken by torture, on an altar of sacrifice that resembles a chopping block. And the crushed grapes, *anawim*, is a word applied to the marginal,

the poor, and the captives. Their blood we drink—as
confession of our disregard for them, and as identity with
them. All our life we are to raise them up in the name of
the One who was crucified once, now, and in the time to
come. There is no other place to be than with the crucified
God who harrows hell. Eucharist is what we do, and that
Eucharist becomes who we are. One who knows, truly
knows, can face straight on the gorgeous chaos, the dread
essence, the terrible beauty, the craving soul of the universe
itself.

Resurrection and the Eucharistic Presence. Instead of
being preoccupied with the nature of or the changing of
the Eucharistic elements, it is promising to identify resurrection as the meaning of Real Presence in the breaking of
bread. This helps to understand better the meaning of both
"resurrection" and "Real Presence."

The Promissorial Eucharist. How helpful it could be
simply to call a moratorium on trying to explain the
Eucharist. Instead, we could see how many of us there are
who could confess simply that "in the Eucharistic event"
Christ does what he promised he would do. Then the
coming forward could be an act of faith, indicating that we
trust the God who makes and keeps promises. As the
writer of the Letter to the Hebrews made clear, "He who
has promised is faithful" (Heb. 10:23). And the Eucharist
is our rehearsal, over and over again, for hearing the
promise.

Eucharist as Spiritual Feeding. The word "Spirit" is
coming into common use among seekers. Thus it may be

possible to speak meaningfully of the Spirit's providing, through the Eucharist, spiritual food and drink for our journey into fullness of life, and likewise, in *viaticum*, for our journey into death.

Scripture as Real Presence. The Catholic Church has finally affirmed Scripture as the *ciborium* (container) of Christ's Presence, and *the* prayer book for all disciples. Are Protestants now able to affirm the Eucharist as the parallel *ciborium* of Christ's Presence? Put another way, can we not agree that God is equally present in Scripture *and* Eucharist?

A Temporal Rather than a Spatial Issue. As we seek more agreement within the Christian family, perhaps most interesting of all issues is to note the shift of focus from the disagreements of the past, which were largely *spatial* ones, to a contemporary focus that is more *temporal*. The primal question is this: *to what degree and in what sense is the past event of Jesus Christ continuous as a present reality?* Eucharist is not understood by the Catholic as simply a re-enactment of a past event. Instead, before his death by crucifixion, Christ presided at the first Eucharist, on Thursday of Holy Week. This event is to be received as an incredible *promise*. Whenever this event is done, over and over again, Christ has promised to be *really* present.

The key to a primary Catholic-Protestant disagreement, therefore, hinges on the understanding of *time*. There is general agreement among Christians that Jesus' redemptive act on the cross and his Resurrection happened "once and for all." We can, however, make significant progress by shifting the question from the spatial to the temporal. If the Christ event is only in the past, the Eucharist cannot be anything more than a remembering of an event that remains past. But what if in the Eucharist we have a convergence of time? Past, present, and future come together in the Eucharist in an event that is *ongoing*, from the beginning of time to its end. At one point *in* time is disclosed the meaning *of* Time. The widely used *Handbook for Today's Catholic* puts it this way: In the Eucharist "both past and future become really present in mystery."[4] Instead of arguing about the nature of the bread in my hand, we might see ourselves as surrounded by Mystery—with the Mysteries of Incarnation, Passion, Death, Resurrection, and Ascension being the pregnant events of time incarnated in space. Figuratively speaking, in the Eucharist all time stops, immersed in space everywhere. And there are three particularly identifiable moments. At the Words of Institution the *Past* becomes Present. At the *epiclesis*, the Holy Spirit becomes the "Real Presence" *now*. And at the "elevation," with the "Great Amen," all is taken into God in foretaste now of the *future*—as the active transubstantiation of all into God. Transubstantiation has been going on since the beginning of time.

In his historical trip to the Holy Land, Pope John Paul II did physically what he had affirmed as spiritually necessary for all us. "It is as if we constantly need to go back and meet in the Upper Room of Holy Thursday, even though our presence together in that place will not be perfect until the obstacles to full ecclesial communion are overcome, and all Christians can gather together in the common celebration of the Eucharist" ("Encyclical: *Ut Unum Sint*"). He has acknowledged that our "intolerant polemics and controversies" are the rendering into "incompatible assertions" what was really simply "two different ways of looking at the same reality." Thus ecumenism must become a central focus, to clear away the rubbish of our needless quarreling. "Ecumenism is not an 'appendix' added to traditional church activity, but an organic part of her life and work, and consequently must pervade all that she is and does." Toward this end, the Pope insists on putting into practice the profound change evident in Vatican II, in which non-Catholics are no longer seen as enemies or strangers but brothers and sisters. In fact, the very expression "separated brethren" is no longer acceptable, replaced by others expressing the deep communion residing in a common Baptism—such as "other Christians" and "Christians of other communities," of "Churches not in community with the Catholic Church." "We now speak of communities," he said, "not denominations."

—‗—

7. Eastern Orthodox Liturgy

In our struggle toward a common meaning of Eucharist at a deeper level, the "Divine Liturgy of the Orthodox Church" (e.g., by John Chrysostom) can be helpful. From the beginning of that liturgy, one realizes that one is participating in a sensual banquet, as both promise and foretaste. The Eternal is not timeless, but is an Eternal Present, marinated in symbolism. The Procession with Gospel and light is the coming of the Savior into the world. The Gospel is the Presence of the Incarnate Lord. The Gloria is the song of the angels. The Old Testament and the Epistle stand for the witness of the Prophets and Apostles. "Alleluia" is *the* human response. With the reading of the Gospel, Part I of the liturgy ends. Part II is the Eucharist. The Holy Table is the sepulcher. The corporal is the linen cloth that wrapped Christ's body. The veil of the paten wrapped his head, and the veil covering both paten and chalice is the stone covering the opening to Christ's tomb. The "Great Entrance" is the way of the cross. Three apses are walled off by the "*iconostasis*" (icon screen), shielding the people from the Mystery, but from it the Saints gaze upon the splendor. One begins to sense that there are two services going on simultaneously—the earthly worship reflecting the majesty of the heavenly one. When the priest brings the paten and chalice to the altar, Christ is buried. The consecration is Christ's resurrection. This event is an orchestration of sight, sound, silence, incense, vestments, gestures, prayers of every kind, singing, chanting, creeds, homily, taste. Here is present the hidden God of unfathomable Mystery. Every sense is stretched, as one is overwhelmed by the spectacle wherein all that exists is penetrated with the Divine. Indeed, the Holy Spirit is life itself.

"Let the heavens be glad, and let the earth rejoice;
let the sea roar, and all that fills it;
let the field exult, and everything in it.
Then shall all the trees of the forest sing for joy."
(Ps. 96:11-12)

Probably nowhere else is there a liturgy that can present so powerfully the glorious meaning of the universe destined for eternity, where creation and resurrection together sing of what has been promised.

—┬—

8. Expanding Sacraments as Sacramentals

We have dealt with the two major sacraments, Baptism and Eucharist. We turn now to the additional five sacraments or ordinances, focusing on their potential for expansion. Secularism has seriously deadened our ability to recognize and function in a world of mystery and symbol. The Church has linked sacraments, or ordinances, to the hinge-points of life's pilgrimage. It is equally important now to encourage an emergence of sacramentals around these hinge-points. The sacraments themselves can be parochial, appearing to be unduly narrowed by the period of history in which they arose. Thus we need to understand sacraments in such a way that their meaning can "overflow," broadening their application. Put another way, where appropriate, the sacraments need to have lay equivalents as sacramentals. Or again, a great need today is for a significant correlation between the Church's sacraments or ordinances, and life in a society largely secular. Such expansion is encouraged, with caution, by the increasing proliferation of unofficial "rites" as alternative ways of entering into the meaning of the Christian life.

Such sacramental openness makes creative contact with what the ecologist Thomas Berry calls "the grand liturgy of the universe." Christian openness embraces as well the feminist experience of the earth as holy and as home. It takes seriously Luther's insistence that in addition to the gathered Church, the "nuclear family" lives as a "domestic church." Applications that come easily to mind include blessing each daily meal; providing water at the entrances and exits of one's home; placing notes on mirrors to remind one of priorities; using a sounding watch to signal a short hourly remembrance; being surrounded by growing things, with cut flowers for special times; appreciating the simple beauty of wood, ceramics, and textiles; recalling one's Baptism when using water, even in washing the dishes; remembering to give simple thanks for the gifts of nature, for color, for sun, for gentle rain; turning off the television when not intentionally watching it, especially when it is merely providing "noise" to suppress loneliness; surrounding oneself with chosen music; dressing according to feast days; doing simple things for others; having a "special" everything—food, chair, room, ice cream, meal, book, music, friend. And in all these things, give thanks.

Henry David Thoreau understood this need for sacralizing one's time and space by calling the "swamp" he entered a *sanctum sanctorium*—a "sacred place." By being rooted in the Christian sacraments, one is able to see with Black Elk that anywhere is the center of the world. To broaden the sacraments so that no one shall go away empty, is to be grasped by Frances de Sales's declaration that when Jesus accomplished our redemption on the day of his passion and death, he knew all of us by name.

The sacraments can be classified according to three types. The sacraments of *initiation* include Baptism, Confirmation, and Eucharist. The sacraments of *vocation* include Matrimony and Ordination. The sacraments of *healing* include Reconciliation ("confession") and Anointing the Sick. Grouped this way, it becomes easier to perceive the sacramentals needed to render broader the Church's blessings.

Eucharist. I remember as a boy a "ritual" we would do when relatives came to visit us. Much like the early Church, mother always invited drop-in relatives to stay for a meal. As we stood around the table, Dad would call one of the men by name, and say: "Would you return thanks, Art?" To "return thanks" is an apt expression for recalling the meaning of "Eucharist." My father was implicitly acting out an awareness of our home as a "domestic church." Just as the Church celebrates every Sunday as a "little Easter," so each meal can be a "little Eucharist"—a remembering in gratitude. Since the Spirit is free to come and go as the Spirit pleases, daily meals in the name of Christ *could* be used by the Spirit to render them as efficacious as the Eucharist itself. Once one becomes established eucharistically, in church and home, the spirit of thankfulness can flow outward, expanding to recognize everything as gift. Short, spontaneous prayers of thankfulness can begin to transform the outer edges of one's spirituality—thanks for family, home, car, employment, friends, strangers, silence, trees, flowers; colors, smells, tastes, touches, sounds; and the music of silence.

Baptism is an indelible marking, and thus to be done only once. But increasingly common now are services that begin with a remembrance of one's baptism, with thankfulness. The only unforgivable sin is the one unrepented. Water gracing the entranceways of church and home, a fountain anywhere, streams and oceans—all of these can bring an ongoing awareness that no matter what, or where, or why, one can be forgiven—washed clean. I find joy in what we do at the monastery. Instead of regarding dishwashing as an annoying chore, it is an invitation to play—with suds and water everywhere, accented with laughter.

Further, the only threat that society or other people have by which to intimidate us is to kill us. How utterly free we are, then, to have death behind us, overcome in Baptism. There is nothing, then, that can block our faithfulness to God. In addition, water recalls for us the sense of immersion—from the seeds of Baptism to the full growth of losing oneself gladly in God. Probably the hardest part of spirituality is remembrance—not forgetting to name the Name.

Confirmation originally was an exercise of the bishop's teaching office. To receive the seal of the Holy Spirit as the beginning of one's adult life required that a person give evidence of having sufficiently matured in wisdom and knowledge of the Christian faith. Confirmation is like experiencing one's personalized Pentecost; and filled by the Spirit within us, the incarnation of God in us is affirmed.

This understanding could make fine contact today with what most excites a teenager—a symbol of the "coming of age." The best that secular culture can do in this regard is to permit a person to get a drivers license at age sixteen, at eighteen one can vote, and at twenty-one one is of "drinking age." But what often goes unrecognized, yet may be most important of all, is one's graduation from high school, with emerging *vocational* plans for dedicating one's life. This could become a rich and important meaning of Confirmation—one that is presently neglected.

I still carry with me leftover feelings from a church camp when I was about sixteen. There was a huge bonfire on the last evening. The leader called for those to come forward who were willing to give themselves to "full-time Christian service." I proudly came forward, intending to become a Christian doctor. I was rather forcefully told that full-time Christian service means "being a 'preacher.'" The Church certainly needs the sacrament of *Ordination* in consecrating persons for the ministries of Word and Table. The gesture of the laying on of hands is a common focus for Catholic and Protestant. It signals choosing from the *laos*, the "people of God," those particularly responsible for the *cleros*, the "inheritance" entrusted to the Church, to be handed on faithfully. But in addition, we desperately need a sacramental consecration of those who have chosen a vocation in which they will function as a Christian. Luther called every believer to be a Christian *in* one's vocation. Thus a banker as a Christian would be honest and fair in practicing this profession.

But Calvin insisted upon more. We are to be a Christian *through* our vocations. In this case, a Christian banker might work to change banking policies so as to eliminate "redlining," thereby helping marginal persons to find humane and affordable housing. What a find it would be to discover a person commissioned to be a *Christian* mechanic—or dentist, or doctor, or mother, or lawyer. Confirmation could mark the transition toward preparation for a Christian vocation. In addition, a sacramental "commissioning" might be done upon licensing, or with readiness to begin practicing one's vocation. One could have one's "nickname" blessed, or be given the name of a saint to be one's model.

The Council of Trent (1545–1563) made authoritative the understanding of Ordination as giving special interior capability. This empowerment is twofold: (1) the ability to offer the sacrifice of the Eucharist (*sacrum facere* = to make holy); and (2) the ability to forgive sin. Catholics and many Protestant churches see ordination as involving both a providential and an ecclesiastical call. But the stance to which Luther came is now more characteristic of the Protestant perspective. Recognizing that in Baptism all believers are made priests, it is for *functional* reasons that a person is designated by the Church to act on behalf of the others. Unlike the Catholic perspective, in which ordination is objective and empowering, the Protestant perspective is more functional, in that it is regarded as being set aside by the Church for a special function on the Church's behalf. The reinstatement of the office of Deacon, in Catholic, Anglican, and some Protestant liturgy, may be the beginning of a sacramental broadening of *call*, both in the Church and in the world.

Marriage is the Church's primary response to sexuality. In this age of rapid change, there is something very special to be said of two persons able to pledge themselves for life, "until death do us part." Yet sexuality, which is the most powerful, beautiful, and yet troublesome dimension of our lives, has been the most difficult aspect of human existence for the Church throughout its history. Part of the tangle is based on "natural law," in which it seemed irrefutable that sex is for the purpose of procreation. Therefore sexual pleasure was denied, even implying that such joy would be sinful. This often led to a sense of guilt surrounding sexuality. At best, sexuality could be affirmed as necessary to alleviate sexual desire and to have children, but theologians seemed unable to see sexual intercourse as a source of grace. Thus Augustine could caution that "a man who is too ardent a lover of his wife is an adulterer, if the pleasure he finds in her is sought for its own sake" (*Against Julian*, 2, 7).

So powerful was sexuality seen to be that it could be regarded as a dangerous force that could destroy society. Only gradually did the Church have anything to do with what initially was regarded as a civil ceremony. It began with a simply blessing after the fact, until by AD 800, liturgical weddings had become normal. But even so, marriage as a sacrament was even longer in coming—for the attitude of Augustine was still strong, that original sin was transmitted from parents to children by sexual intercourse. At the Council of Florence (1439), marriage was included with the six other sacraments.

Only in my time has the Church moved from seeing procreation as the primary, almost exclusive end of marriage, to affirming "mutual love" as equally valid. But the movement here is not yet finished. The Church is much needed especially in our times, because society, in its turn, so misunderstands and abuses sexuality. Some folks insist the Church would just reinstitute a negative approach to sexual practice. Instead, the "thou shall nots" need to emerge from the "thou shalls." The societal game called "dating" is especially in need of serious change. It is an inadequate process for the Christian, and might well be in shambles today. The "working assumption" underlying dating is that the only reason for a relationship between male and female is to find an appropriate spouse for marriage. Thus gender relationships become a means to an end, rather than a way of learning the *intrinsic* quality of such relationships themselves. At its worst, "dating" becomes a game for the sake of seduction and "using" one another—usually with the male being the aggressor, and the female "assigned" the negative task of determining the appropriate level of sexual interaction by imposing restrictions on the male.

It is unfortunate that at this crucial hinge-point in society the Catholic Church unimaginatively declares the sacrament of *Marriage* to be the only real option for laity, or *consecrated celibacy* for those specially called to ordination or to religious orders. The spiritual direction I do is bringing me to question a few of the Church's working assumptions. First, I am convinced that marriage should not be considered normative for everyone, so that if a woman is not married, she is considered "unwanted" and bears the name of "old maid." Or if a man remains single, he must be "gay." Rather, marriage is a "calling," being for some and not for others. Further, for those who marry, I am not convinced that parenthood should be regarded as normative. To be parents is a special calling, so that two persons may be deeply in love and still lacking in capabilities for being good parents. Further, it is unhealthy, and perhaps even un-Christian, to have marriage negate all other friendships of both genders, literally requiring "forsaking all others…."

I wrote an article some years ago proposing a sacramental blessing for commitments of deep friendship. In our mobile society, there is something ennobling about making a sacred promise to "be there" for each other, no matter where either person might end up being. The positive response to this article disclosed a need that we largely ignore. Gender relations are richer and more varied than the options the Church sacramentally blesses. And for those who by nature cannot experience heterosexual intimacy, there is a need to explore serious commitments other than celibacy.

I am encouraged by the current broadening of options in the Catholic Church so that the official Catholic catechism declares that *single life* is "a call no less sacred than the call to be married." [5] On the other hand, married priesthood was permitted until the twelfth century, while at the moment it is still prohibited. Furthermore, the last word on homosexuality has yet to be heard. If it is so that homosexuality is neither learned nor chosen but is congenital, and at least 10 percent of the population is so born, is it not questionable in the name of Christian love for the only options permitted to gays to be celibacy or promiscuity? In a society such as ours, in which promises are increasingly difficult to make and keep, is there not something to be said for a chaste homosexual relationship based on an accountable lifelong promise?

The sacrament of *Reconciliation* was formerly called "confession" or "penance." In our guilt-strewn world, nothing is secretly craved as much as forgiveness. The authority of a priest or minister to forgive sins is one of the most powerful and humbling of gifts. But this dare not mean that only priests can forgive. In a real sense, Baptism includes all of us in "the priesthood of all believers." Being ordained is not so much an exclusivity as it is an assurance that there are persons specially trained and available for this purpose, no matter what time of day or night. Nor does it matter where a person is calling from—a bar or at home with insomnia—a priest is available. How important it is to know that help is available for those ready to have their life transformed, or to turn on end the tragic dead-end of their life.

This need not always be done formally. It may take the form of a kindly smile to a stranger as the down and up escalators pass, sharing hints that one is neither alone nor discounted. In my county, people wave as they pass, especially in pickup trucks. I knew a woman in the inner city who always had the coffeepot on and the kitchen door unlocked, and heard more confessions than any priest. When Paul wrote that "nothing" can separate us from the love of God, he listed everything that could conceivably tempt or characterize us in our time—"hardship, or distress, or persecution, or famine, or nakedness, or peril, or sword" (Rom. 8:35.) There are a host of approaches to healing that rest on the Twelve-Step program, a program thoroughly Christian in its rationale. Especially crucial are Steps Four and Five, in which a full confession of the total failings of one's life moves one toward receiving forgiveness and reconciliation—by a priest or minister.

I came to understand the power of confession and absolution through the ability of little children to come to me as their priest. I have shared tears with them as they remorsefully confessed that they had hurt their dog-friend by pulling her tail. I have experienced the power of the confessional at the end of life as well. I have heard some final confessions on death row. Here are persons hardened by life, some guilty and some innocent, who are able to walk to their death with glowing face—forgiven. So fundamental and sacred is this sacrament that a priest is automatically excommunicated for disclosing anything internal to a confession. In fact, in order to protect women from exploitation, there arose in the Middle Ages a need for anonymity behind a screen. Since Vatican II, one has the option of a face-to-face confession or the anonymity of a confessional booth.

There are four steps to the sacrament of *Reconciliation*. The first is *confession*, as full as one can make. Next is an expression of *repentance* (regret, sorrow). Third is discerning creatively together an accountable way of "undoing" the damage caused. The expression favored here is "*doing satisfaction*." Finally come the words of *absolution*. Gone are the inane "penances" of "saying ten Hail Marys." Rather, for a husband who has been unduly critical of his spouse, his assignment is to heal the hurt by praising her for who she is, along with the gift of a red rose. One of the powerful expansions of this sacrament could be a sacramental to remind one how to say "I'm sorry"—daily. While a lay person cannot give the absolution that a priest is empowered to give, one can listen to "confession" and end it with a prayer for pardon.

Before Vatican II, the Church regarded "last rites" as a sacrament. Now there is a recognition of the narrowness of such an interpretation. Thus, on the one hand, *viaticum* functions as "last rites"—the act of administering the Eucharist to those about to die. On the other hand, the sacrament has become broadened and renamed as the sacrament of *Anointing the Sick*. This anointing with special oil is available to any who ask for it. It is encouraged for those having an operation or those who have significant illness. With its new name, its use could become even broader. There are four kinds of sickness needing healing. The first is the condition of *guilt* coming from a wrong that one has done. For this, the sacrament of Reconciliation is available.

Second is the need for healing from *hurts* perpetrated by others in the past—often resulting in "tapes of inferiority" continuing to play in the present. Most bruises of this kind happen in childhood, when one feels so alone and so vulnerable. For healing, one needs a "cure of memories," in which one relives these painful events, but this time with Jesus as companion and friend, or even as a big brother or sister. For example, with Jesus, one re-experiences the playground where one was always chosen last for "pickup" ball. For many of us there still seem to be chains rattling in the basement after midnight. Yet with Christ, one has the courage to open the door—and often one discovers that what had seemed so ominous was, in fact, mice with megaphones. Anointing is the way in which one dares to believe the divine promise—that because of the Holy Spirit within, we will never again be alone.

A third kind of healing is *physical*. Evidence is considerable that prayer effects physical healing, for which even physicians are inclined to use the word "miraculous." But even those who put limits on the power of the Holy Spirit cannot avoid the very germane fact that the

American Medical Association has ascertained that over 60 percent of the cases seen by doctors are actually "psychosomatic." Some persons wrongly think that this means there is no sickness after all. Wrong. What it means is that over 60 percent of all sickness is *spiritual,* and in need of healing. Karl Menninger wrote a book entitled *Whatever Became of Sin?* His point is that many persons are invaded by guilt, for which the typical psychological advice of "don't guilt-trip yourself" is irrelevant. Only one thing can cure—to confess one's guilt, totally and completely, and receive the unconditional love of God as absolution. This is similar to the first kind of healing, but in this case the spiritual healing brings physical healing as well.

The fourth kind of healing deals with *exorcism.* However one may understand the phenomenon of "evil spirits," it is clear that Jesus cast them out. Further, there are persons today who act as if an evil spirit has taken possession of them, and who really believe that this is so. But Jesus says that casting out "evil spirits" is insufficient, for that only gives space for seven more unclean spirits to take up residence. Thus, central to the Christian faith is the insistence that each person's body is a temple for the Holy Spirit. Therefore, while exorcism is a liturgy of displacement, it also involves *replacement*—for the exorcist works with one or more persons who minister with the one exorcised to help fill spiritually the gaping hole that remains.

In 1179, the Church still spoke of the *Funeral* as a sacrament. We have mentioned that while Protestant funerals tend to be oriented toward comforting the bereaved, the Catholic focus is upon commending the deceased to God. Now, with the funeral understood as a sacramental, the Catholic understanding has expanded to include cremation, pastoral ministry to those with AIDS, and offering an affirmative funeral for those who die by suicide. A positive trend among both Catholics and Protestants is persons helping to plan their own funeral, as well as inviting the family to help prepare and participate in the funeral of a loved one. To the degree that the Church is understood as a World of Sacrament, to that degree can one's life be understood as a sacramental increase within God's world.

—┬—

9. Vows

Part of the deterioration of the moral and spiritual fiber of our society rests in the growing inability of persons to make and keep promises. In fact, society makes it possible to go through much of life without having to make any serious vow. Marriage is the sacrament that comes closest to having a vow as its foundation. Marriage occurs when a couple marries each other by making promises in the presence of Christ and his Church. Yet, half of all marriages in this country end in divorce. In another arena, a record number of personal bankruptcies also gives witness to the diluted meaning of promises. In the monastery, the first or "temporary" vows (after observership, postulancy, and novitiate) used to be for three years. But the Trappist Order is finding that most young people today seem incapable of making promises for as long as three years. Therefore temporary vows now may be made in increments of one year at a time. Even so, the monastery today is called to witness to what life might be like if rooted in promise, lived as discipline, and gathered up into vows. The Trappist Order, to which I am related, has the three traditional vows, plus two more.

Poverty. Ours is a society with little sense of stewardship for the holiness of space, with time bringing a relentless worsening of our earth-home. Ours is the specter of a future filled with garbage, nuclear waste, polluted rivers, and dead Great Lakes—with the prospect of the ocean, Mount Everest, and even outer space becoming cluttered with waste as our signature. To live only with that which is needed, to be a steward sharing what one has, and to own as if possessing nothing—these marks of Christian living call for a radical conversion of our "throw-away" lifestyle. To be able to live in simple material faithfulness, one needs to become "poor of spirit." Probably no society has been as spoiled and pampered as ours, where we are almost incapable of distinguishing between "I want" and "I need." What does it say that two-thirds of the world's population is hungry, even starving, while three-quarters of the population of the United States is "suffering" from obesity?

Chastity. Pornography as voyeurism is now present and available in almost every arena of communication.

Next to professional football, sex has become our leading spectator sport. Monasticism is not an elimination of sexuality, which would be self-defeating. Rather, sexuality is sublimated into acts of deep feeling and beauty. In a society where sex equals a craving for crass release, the monk sees the rhythm of sexuality as one with the rhythm of the universe. The monastery calls into judgment the tendency of weddings to be a give-away of the bride; and in the time of the Bible and perhaps our own, adultery was forbidden because it violated a man's property rights. In strong contrast, chastity, religiously held, means to love nonexclusively and nonpossessively.

Obedience. Scripture speaks much of doing the will of God. Perhaps obedience, as we noted, is best understood as faithfulness to the divine *yearning.* In our society, discipline is hardly valued—except for athletes as a means to mercenary ends, or the Marines as a means to violent ends. This is because our societal "values" are instrumental rather than intrinsic. What our society desperately needs to learn, although the Church is not yet able fully to model it, is how to live based on *intrinsic* values—life lived for its own sake. This applies directly to the vow of obedience. Most people today will "submit" to being obedient only if it is clearly worthwhile for them. As a vow, however, one learns to be obedient for the sake of obedience. Thus the monk might be told to scrub a floor that has already been mopped twice, for no other reason than to be obedient. Many of our "addictions" today result from our never having mastered the discipline of self-control. For the monk, the purpose of obeying is for the sake of obeying the God who is loved for the sake of loving.

Stability. In a country where most persons move once every three years, the monk is an oddity. He or she takes a lifelong vow to a particular monastery and to the particular place on earth where it stands. There is something awesome about spending one's whole life loving into being the monastic 1,000 acres or so—to know each creature and every growth as befits a gardener in love with creation. Still profound is the advice by a desert father to a disciple tempted to travel in search of "the answer." "Go into your cell," said the saint, "and it will teach you everything."

Conversatio morum. (*conversion of life style*) This Vow of vows is a commitment to render one's life pilgrimage an ongoing "*growth in grace*." There is no salvation outside the Church, if we mean that outside it one cannot accomplish the fullness of being alone. We deeply need those who love us unconditionally, help us discern the next step in our pilgrimage, hold us accountable to a faithfulness in that journey, and for this purpose feed us by Word and Sacrament. The goal is purity of motivation, in which everything follows from one's unconditional relation to God alone. The Trappist technically does not make the vows of poverty and chastity, understanding these as implicit within the other vows. Some interpreters see them as implied in obedience, others in stability, and still others, of which I am one, see these as implicit for the monk in "*conversatio morum.*" And just as the monk makes vows rooted in a Rule (e.g., Saint Benedict's *Rule for*

Monasteries), so each Christian needs to be so rooted in a rule of one's own distilling, within a context of one's own effective accountability.

I am part of a special group of four persons, from different parts of the country. Yearly, we meet for three days of mutual spiritual direction. The assignment they gave me for this year has been quite challenging. Since I embrace so deeply the event of Eucharist, I am to explore what it might mean for me to "be" Eucharist. *It occurs to me as I write that, unknowingly, this might be what this book has really been about.*

The monastery as a model is a haunting place. In risking a visit here, one might discover that to live in the tranquillity of the Spirit's Presence entails abandoning one's ego, giving up one's ambition, distributing one's possessions, and discarding every temptation to self-aggrandizement. A Christian pilgrimage of "letting go" means rooting out the "rewards" that society has ingrained in us to want. It means giving up all this in order to chant and pray, day after day, and year after year—linking mystically all persons as part of "Holy Mother Church"—which is yet fully to be. In spite of scandals and divisions, the Church at prayer quietly continues on—committed to trust the promises of the God whom we praise. Yet what is surrendered is so small, when seen in terms of the peace and joy of living in the Presence. To become free is to become lost in creativity for God, and gladly to shoulder in fact and in symbol the suffering of the world, interceding for everything and everyone. Christianity is an invitation to be part of the great cloud of witnesses spanning the earth with the incredible vision to which God is calling us as co-creators—where the drive finally is not so much to belong as to unleash, form, mold, penetrate, and soar in restless craving. This is the Eucharistic transformation of creation.

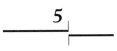

5

THE SHAPING OF SPACE AS BEAUTY

"One thing have I asked of the LORD, *that will I seek after:*
to live in the house of the LORD *all the days of my life,*
to behold the beauty of the LORD, *and to inquire in*
[God's] temple." (Ps. 27:4)

"[God] has made everything beautiful in its time."
(Eccles. 3:11 RSV)

To live with Presence at the center leads almost inevitably
to a life of creativity of all kinds—each in its own way
creating in space expressions of the incredible vision
birthed by the Christ event. "The Revelation of John"
strains every sense—sight, sound, touch, smell, and taste—
in glorifying the Alpha and the Omega. The One with the
keys to unlock even death has a face as the shining sun,
star-studded crown, hair white as wool, eyes of flaming fire,
feet of bronze, rainbows as halos, legs like pillars of fire,
and clothed with clouds. The sound is of many waters, of
lightning, of voices as peals of thunder with hail, and
harps, flutes, and trumpets, singing eagles, music sung by
thousands upon thousands—and silence. Eating is from

fruit trees of all kinds, with wine, grapes, bread, and hidden manna. The sights are of heaven and earth, temples and valleys, hot and cold, thrones of flashing light, falling stars, floods, and caves with torches. The surroundings are of rivers, fountains, burning mountains, a sea of glass. The smells are of smoke and incense. The richness is of colorful horses, bloody moons, green grass, bright linen—gold, frankincense, and myrrh. The creating is of wood and glass and stone, of jasper, sapphire, emerald, and crystal. And because all is for the marriage banquet of the Lamb, the promise is clear—that "Nothing accursed will be found there any more" (Rev. 22:3).

Over the years I have gained a special appreciation for those called to create at the intersections of beauty, who establish an environment for multiple senses through multiple arts. It began for me when I recognized music as deep theology, such as the music of Johann Sebastian Bach, who interpreted the Trinity through fugue. The composer Richard Wagner had an inclusive vision, fusing the auditory and the visual in a "world" happening. Architect Frank Lloyd Wright has particularly intrigued me. Not only was he a genius of formed space, but he perceived profoundly how creativity emerges best from an environment of beauty. Thus his studios and architectural school, at both winter and summer Taliesen settings, were atmospheres of music, vistas, artifacts, clothing, food, aromas—part of what he called "vision." Share with me my excitement in discovering for the first time that this morning as I write into my computer, it is able to gift me with Vivaldi.

I suspect that each of us has secret longings, some of which can be touched by the question, "What would you be if you had life to live a second time around?" I know one of my colleagues would opt to be a major league pitcher, another a professional jazz pianist. I? To be an architect. I don't mean instead of the first go-round, but in addition, as an expansion of my present life as a theologian. For me the two callings form a whole. In fact, their uniting only makes more intentional what it means to be a Christian. We are called to sacralize our spaces into the fullness of life with which we are gifted daily. How blessed it would be, then, to have the talent to be professionally a theological shaper of beauty. But what a terrifying responsibility that would be, for what one designs architecturally has enormous power to nurture or wound the human soul, to deaden or render alive the human spirit in all its doing and relating and being.

—⊦—

1. Beauty as Transfiguration

Beauty is a word seldom heard today. Efficient, productive, useful—yes. But beautiful? Or if it is heard, images are of Breck shampoo or nominations for being the "ultimate driving machine." Yet I remember a suggestion a researcher made concerning the multiple problems surrounding present-day sexuality, personally and socially. "I propose that we have a moratorium on all talk about sexual technique, even about sexual ethics. Instead, let our focus be on beauty." Such choreographing of our living and its spaces has been a significant aspect of the Church's thinking and doing through the ages. As early as Augustine, we find beauty to be a central category even for understanding God.

> Beautiful is God, the world with God…. He is beautiful in heaven, beautiful on earth; beautiful in the womb, beautiful in his parents' arms, beautiful in his miracles, beautiful in his sufferings; beautiful in inviting to life, beautiful in not worrying about death, beautiful in giving up his life, and beautiful in taking it up again; he is beautiful on the cross, beautiful in the tomb, beautiful in heaven. Listen to the song with understanding, and let not the weakness of the flesh distract your eyes from the splendor of [God's] beauty. [1]

But while persons might notice something "attractive," or observe someone who is "pretty," beauty is an intrinsic quality for which our senses must be formed if we are to participate deeply. Otherwise Picasso's women don't look like women, and Van Gogh's stars don't look like stars. Particularly in need of formation are eyes capable of being grasped not simply by the beauty of the extraordinary, but of the ordinary. The profound contemporary sin is "taking things for granted," diluting our ability to enter into the beauty of anything, our soul closed even to the absolute wonder simply in "being." Formation is the Church's business, as it were. In her, and through her, we can be shaped to know beauty as the "liturgy" of things, as a whole and each, in their "thusness" and "suchness" and "eachness." As mentioned in Chapter Three, the poet Kathleen Norris came to recognize this through her experiences at a monastery. Here she was grasped by how the monks weave ceremony through every mundane part of life—how one eats, dresses, works, even how one enters the sacred spaces.

I remember my disappointment some years ago after spending an evening in the home of an accomplished conductor. I eagerly anticipated the prospect of listening to music with him, discussing our favorite loves. To my shock, his record collection consisted only of his own performances—and our only conversation was of his accomplishments. The accolades of conducting had crowded out in him the love of beauty for its own sake— which apparently he once had had. While many know the notes, I have come to know that there are far fewer who know the music.

Beauty has to do with our whole disposition. I sense this as a predilection that begins as I drive down the lane to the monastery after having been away too long. Even the gravel's crackle is too loud, and the breeze through my hair becomes an event. Swept by the silence, I am enabled to hear afresh—the sounds of cattle far over the valley, wind through the pines, even the feathers of birds' wings gliding the air, and coyotes through the moon-shadows of the night. Sacred. What a tragedy that such living is lost for most folks, for sacredness untrained is preciousness lost. Even Saint John of the Cross, sometimes criticized for his preoccupation with the "dark night of the senses," took his monks into the mountains regularly to talk of God's beauty. Likewise, his spiritual friend Saint Teresa of Avila, wisely instructed her nuns: "I do not require of you to form great and serious consideration in your thinking. I require of you only to look." Looking is the focus-point of loving. And for her, looking meant being "the joyous nun," singing and dancing and playing the flute before her God. Such seeing requires only the simplicity of standing at attention before life: tasting the dark, truly hearing a bird, entering into the candlelight, savoring a glass of wine, stroking the silky-smoothness of a piece of walnut, or being surrounded by a midnight gentleness. "There lives the dearest freshness of deep-down things," insisted the poet Gerard Manley Hopkins.[2] Andre Louf sees monks as a witness for everyone. The purpose of monastic withdrawal is to go into training. Here monks are touched by "the immense and ever-active yearning which runs through all creation," and hear the call "to bring about the transformation of the universe." [3]

Although the Church has not always been careful in speaking of such matters, to be consistent with the gospel, the Christian must never be blind to creation—nor reject it for some seemingly "spiritual" reason. How presumptuous to discount the whole of God's universe for the sake of some inner experience or presence. Rather, what needs to be rejected is the temptation to *possess* creation, thereby abusing it of all intrinsic meaning. We obliterate creation's sacredness by reducing it to self-serving utility—for if it is always a means, it is no longer an end. But once one sees God definitively in the Christ event as the divine Incarnation, one can discern God everywhere, in every nook and cranny. Just as the Church is not an "it," so neither is Nature. Both need to be heralded and embraced lovingly as "She"—as the body of Christ and the temple of God.

Paul said of Christ that everywhere he is always "Yes!" (2 Cor. 1:19-20). Nothing is outside the parameters of God—and Paul makes sure we understand this. "[Neither] height, nor depth, nor anything else in all creation, will be able to separate us from the love of God in Christ Jesus our Lord" (Rom. 8:39). Christianity must be a definitive "Yes" to all of space—even of "hell." Nothing declares this inclusiveness better than early liturgical expressions for the final days of Lent, the bleakest season of the Church year. In a Good Friday hymn written sometime before the eleventh century, we chant of the consequences of crucifixion for all creation, as "the sun began to fail in force, the silent stars began to weep." But then "above the darkened earth was raised the Son whose arms like wings were poised." All depends on that sacred moment, for it marks the death or the re-creation of all things. "As on the first of days, [Christ] breathed forth his Spirit on the void, and a new creation began, one formed by grace to make us free."[4] Another ancient passiontide hymn proclaims the significance of Christ climbing "the altar of the Cross":

> *Earth and ocean, stars, creation*
> *Washed and cleansed by this pure flood!*
> *Worthy to prepare a harbor,*
> *Pilot home the shipwrecked world.*[5]

Equally powerful is an ancient undated homily for Holy Saturday, portraying the "harrowing of hell." Between Good Friday and Easter, Jesus "descended into hell," as the creed states. Intriguing is why he went—in search of Adam, as if for a lost sheep. There he embraces and invites "all who are held in bondage to come forth, all who are in darkness to be enlightened, all who are sleeping to arise." Christ's reason is clear: "I did not create you to be held a prisoner in hell." The sacred vision, then, is even for Adam, in spite of the fact that it was he who was responsible for the condemnation of creation to bondage and decay: "The bridal chamber is adorned, the banquet is ready, the eternal dwelling places are prepared, the treasure houses of all good things lie open."[6] The table in the desert is set, and all of space around it is rendered sacred for the celebration. From this vision, Gerald Manley Hopkins insists: "The world is charged with the grandeur of God."[7]

I remember from her journals the image of May Sarton surviving the severe New England winters. When the wind drove temperatures below zero, her response was

one of "belligerent beauty." She braved the storm in order
to have cut flowers on her table.

—⌐—

2. My Own Efforts at Sacralizing Space

A significant mark in my own spiritual pilgrimage has to
do with space, in two senses. First, I came to a point in my
life where I needed to go somewhere else as "a place
apart"—to a sacred space where people were being, in a
sense far different from anything I had known up to that
point. My choice was a monastery, removed from "modern"
living. Ongoing immersions in the sacredness of monastic
space and time drew me first to sacralize the times of my
living.[8] More slowly was I lured into the reality of sacred
space. This urge was tentative—at first eliminating plastic
from my life, simplifying furnishings, eliminating clutter,
naming a sacred corner, minimizing television use, discarding "gadgets," frequenting thrift stores, and sharing my
possessions. But in time, the yearning became unqualified—
to bring closure to a season of my life, and to prepare for a
new one through early retirement. This next season
dawned with a firm desire to live life in an intentionally
sacred space.

As a child of the Great Depression, I learned how to
"make do," to charge nothing, and to make almost anything
out of orange crates. In Appalachia we were ecological
before the word was invented. But at the beginning of this

new season in my life, guilt nibbled at me as a daily companion. I shared anything I had, and gave away freely. No problem. But there was no way that I could justify spending money on myself—especially in building an environment of beauty—when so many persons had nothing to call home. Mind you, what was happening had nothing to do with *Better Homes and Gardens*. All I knew was that something happening in me had something to do with shaping and forming, much as music and dance relate to being alive.

My compromise may sound strange to others, but it was my only option. I took a "side-job" tearing down condemned tenements near where I lived in the inner-city. My pay was in "resurrected" lumber, and all the bent nails I could salvage. Every weekend I brought a pickup-truck load of reclaimed boards to a plot of land I bought in a cedar forest on the shore of a flood-control lake. For two gracious days each week the straightening of nails was my mantra. And what in due time was birthed was a hermitage of beauty, built by my own hands out of my own dreams. In its own way, it is an ecological atonement for the waste of our society and for the prostitution of "beauty." Actually, the hermitage built itself, and I was midwife. This hermitage is where I live now as a hermit and monk—also helping the poor, and doing spiritual direction in this space of peace for those seeking sacred space and time in their own lives.

The main pillars for my hermitage came from a house of prostitution on Twelfth Street. The basic color is that of barn wood, painted by the rain and wind. The interior is one of open and overlapping spaces, with functions suggested by different levels and angles rather than walls. The cathedral ceiling points toward a sleeping loft, cuddled as intimate space near the eaves. Large windows everywhere are channels of light, arranged so as to bring inside the daily rhythms of the sun. The perimeters between inside and outside are intentionally blurred. While there is an overall sense of simplicity, the wall spaces are for me rich with "treasures"—some given, some hand-carved, and some woven. The visual focus is a relief marquetry made on the last night of my initial construction—when I swept the floor, and arranged the final scraps on a four-by-four-foot plywood backing, in a form that only later was visible as a bursting sun. There are almost no ninety-degree angles in the whole home, nothing predictable, inclining more to the serendipitous and unexpected. Angles and steps and a circular staircase provide shadows and invitations and encouragements to guess and to explore.

The materials are fiber, wood, glass, hanging beads, ceramics, statuary, paintings, sketches—of every color. There is a library, but books and reading materials appear on most flat surfaces as invitations to read. There is music for the asking, and a keyboard for the adventuresome. Comfortable flea-market chairs are arranged to summon sharing. Whatever, there is space for "it"—for play, worship, writing, meditation, eating, praying, creating, sleeping—for being and for praising. There are invitations to fun—Mickey Mouse is prince here. Clay and chalk are present, and a shop with fun tools invites one at least to smell the varied woods. There is an open chapel and tabernacle, with a perpetual candle that especially at night flickers as sign of the divine Presence. Plants are everywhere—hanging, sitting, climbing, and multiplying, with greenhouse alcoves off the loft. A circular fireplace is at the physical center, its flames visible from each part of the main floor, and reflected for the loft from varied windows. Standing in the loft one can look down through the banisters at a vista to the sitting area, featuring a monk's desk overlooking the lake. Or one can look out through the wall of windows above the trees, open at night to the stars, and to the moon as it makes its nightly trek. Windows and doors, open in all directions, make the lake breezes frequent companions, welcoming the holy times of sunrise and sunset.

Outside are intersecting decks down through the cedar grove to the lake, as stations of the cross—illumined for evening walks, glorious in the spring when the whippoorwills are lavish with their vesper alleluias. On one end of the expansive deck is an outdoor sunken hexagon where I celebrate daily Eucharist when the weather invites such celebration, with animals and owls providing the visible congregation. The land overlooks the lake, with rocks and paths through the woods and provisions for swimming and paddling. The exterior is an intentional feast of the senses: a pool with waterfall, the flash of orange fish, oak and walnut firewood, bird feeder, sundial, and birdbath. Intermingling with the bird sounds are wind chimes cut for the musical intervals of Gregorian chant. The entry is a gateway within a fence open enough to invite, shielded enough to entice. One passes under a tower whose doorbell is a church bell, with pull rope The double gate needs only a gentle push for entry; but once inside, the handle for going out is in the shape of the world, one half on each door. To leave, one must grasp that world and pull it toward one's heart. Above all, this is a place of peace—a deep peace in which an alternative beauty breeds gentleness and nonviolence, and a sense of "coming home."

Those who know me well say, "This hermitage is *you*." Yes—and yet it speaks deeply to others as well. There seems to be a multiple richness here that can touch a diversity of persons—except, perhaps, those for whom noise, incessant activity, and perennial television are "necessary" ways of life. Yet even these seem to sense that in coming here they need to sacralize their *time* into a pilgrimage. And for those able to go deeper, this hermitage seems to evoke most a yearning to sanctify the *space* of their existing. Whatever form that may take, it is usually a countercultural expression, as these persons attempt to counter the lethal shapes and patterns that characterize the interior and exterior spaces of modern living.

—⊢—

3. The Church and Theological Space

Throughout its history, the Church has been a patron of beauty. It is as if quite early she sensed a vision and mission to intersect space and time with the arts as creative ventures in sacredness. As the disciples struggled to grasp the life transformation entailed by Jesus' Resurrection, they moved into the upper room, setting up residency there in order to be formed. Since Vatican II the Church has undergone a resurgence of interest and experimentation in creating sacred space that has the power to form one's spirit.

Architect Duncan Stroik, however, Associate Professor of Architecture at Notre Dame and editor of *Sacred Architecture*, recently criticized an important document in this resurgence: "Environment and Art in Catholic Worship," from 1978.[9] This booklet, produced by the U.S. Bishops' "Committee on the Liturgy," nonetheless needs to be appreciated as Vatican II's response to the previous period, in which "beauty" tended to be seen as decoration, adornment, addition, and imposition. The necessary response was for "simplicity" without "starkness," in which beauty is intrinsic to function well formed. Nonetheless, Stroik has a point in claiming that this affirmation of a "low church style" has tended to become "a paean to modernist abstraction." As a result, he observes, tabernacles are removed from sanctuaries, religious imagery disappears, and the resulting feel is of the puritanical. Quite interesting is what he names as inexcusably missing. Intent as liturgical reformers have been upon requiring "theatre-shaped interiors," "there are no columns to sit next to, no shadows for a penitent to kneel in, no places for private devotion, no mystery and no images of the heavenly host." I agree, for it is significant that caves fascinate even children, and each child finds a little corner in which to feel the quietness. Everything needs darkness, especially today when the values of the night are deprived us by a violent society in which darkness is synonymous with fear.

But Stroik's conflict ultimately is not architectural, but theological. What he laments, actually, is that contemporary church architecture tends to express only *one* "theological world." [10] There is no sacred space for those who are fed deeply by mystery, or those whose search for meaning is a quest for forgiveness, or others still whose need is for the warmth of intimacy, or those whose healing craves the solace of darkness. In other words, the Christian approach to beauty must involve the richness, ingenuity, and diversity of sacredness that is often as missing in our churches as it is in society itself.

Promising is the work of Franciscan Sister Ann Rehrauer, associate director of the U.S. Bishops' Secretariat for the Liturgy and staff coordinator for the new art and environment text being created. The beginning will be a basic theological framework, followed by a consideration of the nature of liturgy itself. Then, emerging from this foundation, will be the exploration of how art and architecture "should express what happens there." [11] The result will be beauty understood in terms of a rich diversity of meaning.

Space-time defines all that we know. We are surrounded and invaded by the three dimensions of space, plus time. Fascinating are two contrasting but related definitions of space: Space is the lapse of time between two points—

thereby rendering space the environment of pilgrimage. Secondly, space is the shaped volume defined by form—rendering architecture the art of sacralizing space as pilgrimage. For years I have taught theology in terms of an action-reflection methodology. One of my favorite adventures in doing this involves taking students on an "architectural immersion." My intent is for them to participate in contrasting "theological worlds," inside and out. While classical theological texts can help them articulate examples of theological diversity, it is at least as important for them to experience the "feel" of life from within the sacred space of diverse theological "residents." Beauty entails an immersion in shaped meaning, rich because it has the ability to open us to its different kinds and forms. Thus, for example, some persons are claimed by a Cezanne painting more than a Renoir, others by a Matisse over a Van Gogh, or a Rembrandt more than an El Greco. Which is the better painter? At the level of the brilliance of such artists, the question is meaningless. We are dealing with different visions, with the different World in which each painter lives and moves and has his or her creative being. Such paintings are contrasting entries into reality, invitations from residents of different "theological Worlds." Together they express the richness and fullness of existence, at the same time capturing contrasting residencies from which one is invited to see the whole.

This is as one should expect, for the world created by each of the four Gospel writers cannot be condensed into one. Each world is unique, emerging from the interaction between that author's need and the resolution of that need which was experienced through the Christ event. Each pilgrimage shaped and formed the special space of each gospel world.

So it is that each church building, for better or worse, invites one into a particular version of a Gospel World. Thus nowhere is the power of diverse beauty more apparent than in the history of the Church's immersion in architecture.

But so often, the deadness of many church spaces results from structures informed by a least common denominator of sentiment and memory, reflecting the tasteless commonality of generic Christianity. "What should a church look like?" is a question whose answer is usually little more than an association drawn from one's childhood upbringing. Instead, one can awaken to the powerful differences that exist in Christianity's rich spiritual heritage, by actually walking into alternative expressions of beauty. Let me take you on such an "architectural immersion."

New England Colonial. My students and I begin with the seminary chapel itself, typical of a host of church buildings throughout this country. Originally it was designed as a *house*—a house of God. Thus in structure it resembles the homes of New England towns themselves. Placed on the opposite side of the green from the town hall, which was the "house" of government, New England churches expressed an architecture of function—the significance of a building depended on what the people did in each particular space. But all the spaces reflected the Enlightenment, a time vigorously informed by rationality. Consequently, the church space is characterized by rational lines and composed of geometric shapes, with clear glass windows open to the knowable world, and brightly lit by white painted interiors and furnishings. The pulpit is central, because from it the sermon was delivered as the means of instruction and learning. The impact is that the interior is oral rather than visual.

Ironically, while the original design fostered a harsh declaration of divine judgment and the call to conversion, this design today has become characteristic of Protestant liberalism. Often now the color is a gentle blue coloring, providing a restful atmosphere in which life as understandable is explained and illustrated. What you see is what you get in such space, with no mystery and little intentional symbolism. Much is left to the mind, and little to the imagination. When one leaves this sacred space, it is with an awareness of why this is a favorite design for suburban locations, for it witnesses to a neighborhood harmony between church home and domestic home.

Swope Park United Methodist. The pastor of this church when it was designed had been a seminary professor, committed to a "neoorthodox" theology. The meaning of this space is almost all internal. Immediately upon entering it, one is struck by the fact that no light comes from the outside world, with the high windows on either side wall being tints of black and blue. On the one side, the "bruised" glass permits patches of lighter tone in which are symbols of Holy Week, such as sponge, nails, and ladder. On the other side, several "bullets" of color in the glass suggest the work of the Church, as in preaching, Baptism, and Holy Communion. An extremely long "pilgrimage" aisle stretches from back to front, powerfully suggesting the long trek of history toward its climax. And that climax is clear, for in the far front is a wooden cross firmly anchored "for the ages" in solid rock. A pulpit is slightly to one side, but clearly dominant for preaching the Word. The length of the whole suggests the old adage that the proper form of proclamation is the shout. Surprising, in contrast, is the smallness and portability of the baptistery. But, appropriate to this setting, the baptismal bowl is a metal sphere anchored within wooden "trinitarian" legs. The handle on the top of the sphere is a cross with which half the "world" can be lifted, almost like the firmament, opening the world to the baptismal water—recalling the water on which Genesis understood the world to float. This same sphere-and-cross motif is etched into each pew, branding the congregation to go out into the world to baptize in Christ's name.

The ceiling, made of spun glass, suggests clouds. At the middle point it almost appears to sag from the eager weight of whatever is above it. It seems to float low over the congregation, much as the cloud that led Israel. Light streams in from around the edges of the cloud ceiling, and erratically sunken holes break the surface with unpredictable hints of light "from above." As one walks the center aisle, one senses in the whole an Advent flavor, one of heavy yearning for the inbreaking of the divine. And so it is that one's visual attention is drawn to the raised altar in the front, where overhead the only natural light breaks through and streams downward onto it. Electrical lighting guarantees that one can witness this inbreaking even at night. On either side of the altar is a latticework of brick and spaces, with shadows hinting a color from hidden windows. Appropriately, then, the whole building is so situated so that on December 25 the rock front is arrayed in magnificent color from the hidden source.

As one exits down the long aisle, the rear doors of solid wood hide the center of the huge stained-glass window, hinted at by clear glass around them. When the double doors open, the narthex is overpowered by a soaring, flame-and-blood-colored, two-story, stained-glass portrait of the Trinity. A stern Almighty figure sits on the throne of judgment, holding the cross, with a dove as his feet. When one leaves this sacred space, it is with a sense of the power of gathered memory.

Saint Francis Xavier Roman Catholic Church. The symbolism involved here is declarative more than evocative. The huge whole is an explicit ark structure, with a saintly figure on the bow of this ship of salvation. Seen from above, it suggests a fish—such as Jonah's whale as a symbol of Christ in the tomb, and the ancient code figure by which Christians identified themselves. The interior walls are high, closed to the outside. Yet there is translucent yellow light giving the interior its illuminated appearance. The ceiling is high, disappearing into a painted surface that suggests a vagina, from which one can sense a womblike anatomy for Holy Mother Church. Although not actually Gothic, the upward solidity is not unlike an ancient cathedral as the eternal center within a fortressed city. The "crucifix" high over the altar is created of light space-age materials, and, without any cross, Jesus soars upward, free of earthly constraints. When one leaves this sacred space it is with a sense that what is promised is yet to be.

B'Nai Jehudah Temple. From a distance, this Jewish temple suggests a huge tent. While it is off-center, there is a huge center pole on which the whole structure rests. As one approaches, the image becomes more that of a mountain, with a pathlike concrete spiral suggesting a path ascending to the top. Upon entering, one is immediately claimed by "blueness." One whole roof-side is a mottled glass of peaceful blue, in the same color of quiet receptivity as the comfortably upholstered seats. Lest there be any lingering doubt that we are at the foot of Mount Sinai, on the centering pillar hangs a huge arklike box in the shape of the two tablets of the Ten Commandments. We are at the structural and spiritual center. With the touch of a hidden button, the ark slowly opens, as if by Invisible Hands, as interior lights make startling the silver and gold covers of the hand-scribed Torah. In the center of the large front platform is a blue upholstered table upon which to unroll the scrolls, with a desk from which the Law is expounded. Primary is the Law, secondary is the commentary. The spiral of ascending walls opposite the translucent roof-side are covered in cloth, suggesting the interior of a tent—as in Israel's "tent of meeting" in which God appeared to Moses. The tent flaps, as it were, in the rear can be drawn back, opening not only additional seating space but also an area where festive meals can blend with the worship. When one leaves this sacred space it is with a sense of a divine disclosure calling us to listen.

Longview United Methodist. From the outside, this church is unimposing. In fact, were it not for a slight cross, it could be mistaken for a circular apartment house that fits well into its residential surroundings. The simple entrance opens immediately into an extensive foyer, which resembles an interior flagstone patio. It is well lighted, with large windows opening out onto the rear lawn. One senses quickly that this space is a place of meeting and sharing and happy, noisy hospitality. So it is on Sundays, with coffee and donuts being the prelude for entering the designated worship space, only steps away. This hospitality patio circles around the worship area's curved wall of sliding glass doors, providing easy entrance or exit for gathered worship. It also invites easy interchange between the two spaces—as organ or guitar music from one area filters into the other, and coffee cups and hymnbooks often share a common space.

The worship space proper is circular, sloping down from all sides into the hollowed-out earth. In its center is a Communion Table, its shape suggesting a family dining room table. There are kneeling rails on all four sides, with all the aisles leading down to that point. All the pews are angled, some of them twice, so that one is able to see some persons who are sitting in one's own pew. Surprisingly, this church seats approximately the same number of worshipers as Swope Park, but instead of being thirty pews in length, there are only ten in any of the radiating directions. This gives to the whole the intimacy of a family setting, reminiscent of the worship of the early Church in homes. In this setting, I personally feel the human passion to touch.

Of architectural necessity, preaching here was meant to be a quiet sharing, with the sermon a disposable art form, and communion like sharing from a family album. Here one senses how much preaching can be a deeply social, even dialogical event. The intent was for the pastor to share from the table, thereby being the lowest of the worshipers, not only quite visible but clearly part of the congregation. The primary color is a blood-red carpeting covering the whole area.

What I have described is the architectural creation at its best, but there are unfortunate compromises. For one, the choir insisted on having its own space—so that the "in-the-round" space became a "two-thirds in the round," with the choir functioning to give the whole space now a front and a back, itself becoming almost the focus of worship. Likewise the pastor, becoming squeamish (or ego-threatened) about sharing the Word from different sides of the Table, had a pulpit installed as a large stationary fixture, requiring, ironically, that it be placed high above a hot-air delivery system blowing outward toward the congregation. While there is a large flat skylight over the Table, it is hardly recognized, for light streams in from the eight surrounding clear-glass floor-to-ceiling windows. Through these the world inside the church is unabashedly visible to the whole neighborhood, while the traffic of the main streets outside is part of the interior rhythm.

One evening in that space I came to realize that sacred space often seems to happen of itself as much as it may occur through design. I was to dialogue with the pastor about the theological meaning of the sacred space that the congregation had created. In sharing with them what I saw to be the significance of the plain-glass exchange between interior and exterior worlds, the pastor sheepishly confessed: "We weren't able to afford the stained glass." And in waxing eloquently about how the red carpet suggested that a worshiping community is rooted in the blood of the saints, the chairperson of the building committee was correspondingly sheepish: "The red was on sale." Be that as it may, when one leaves this sacred space one knows that one belongs, as an important and integral part of God's family.

The Temple of the RLDS World Headquarters. This new church of the Reorganized Church of Jesus Christ of Latter-Day Saints rises majestically above the urban horizon of Independence, Missouri, visible miles away. The exterior shape is that of a conch shell, spiraling from a heavy foundation to a thin point high in the sky. The ascending movement itself is awesome. The front doors, over which hangs a fifty-foot cross, enter into a brightly lit, glass-roofed foyer with information counters, almost suggesting a music hall. From there, the suggested entrance to the sacred space is along a "Worshiper's Path." Its beginning is through etched glass doors, suggesting the grove of trees where the Mormon founder received his initial vision. From there the circular path mounts slowly upward, the stonelike walls on either side suggesting the catacombs of the early Christian worshipers. Along the way there are benches and respites of focus, and at the halfway point one is required to pass through the shadow of the cross. At this point one is able to see the end, an apparent wall that increasingly becomes visible as an overflowing fountain—with baptismal words etched into the marble.

Just as in Baptism one's life takes an abrupt turn, so here one passes out under a low-slung balcony ceiling and, with one more step, the space suddenly shoots upward, with an infinite flourish. Ceilings upon ceilings spiral upward, as if formed by the interior of nature's shell—bathed in natural but hidden light, filtering the spiral motion upward. One is drawn in, as the tints of color move from white to beige to pink, as does the interior of a shell itself. Most persons find themselves walking slowly until they find the exact center point, where with head bent totally backward they are directly under the disappearing point. The spatial contours and one's physical postures keep intermixing, as the amazing interior gives one deep feel for one's own vast interiority. The patterned pipes of a tracker organ, itself a fine work of art, promise the intersection of sight and sound. The only explicit symbol in the entire space is a huge Communion Table—whose weight of half a ton insists upon the centrality of its immovable purpose.

When it is time to leave, one discovers through the architecture that when one truly worships, one never leaves the way one enters. The exits are at a higher level than the entrance, and one moves through them into an area overwhelmed by a green-and-yellow stained-glass wall, which portrays with wheat and rice the call for workers in the worldwide harvest. The huge bronze doors have cast into them a child playing with the lion and the viper as the portrait of peace. And when the doors are swung open, the vista is of wide steps leading downward into a huge piazza. Constructed of colored brick, it forms a map of the world—through which one has no option but to walk. When one leaves this sacred space, one feels a deep and redeemed interiority drawn outward into mission.

The Community Christian Church. Some time ago this congregation hired Frank Lloyd Wright as architect. To me, the sacred space was better in its original blueprints than in its final form, after bitter conflicts caused compromises to be made. While intended to be the first modern structure to stand on stilts, the present door is unimpressive. Yet I find the dark interior to be intriguing, where strange low-ceiling corridors, shaped with strange angles, eventually disclose sideward-slanted steps moving upward. It is as if I have to find my own way, with the risk of becoming lost. In fact, I have the feeling that it might be best to walk slightly stooped, for safety's sake. For a short time there is a hint, absent in almost all other church spaces I know, of the mystery of the catacombs, which characterized the earliest Christian worship. At the top of the stairs, a low-slung balcony without apparent support shields me from having any vantage point—until suddenly one passes into an interior ablaze with natural light. Through the spaces of a rhythmically sculptured ceiling I see up through the glass in the roof—forever. And lest the aspiration of the human call be forgotten, at night two beams of light pierce through the latticed ceiling, their mile-long rays intersecting at a pinnacle high above the city. Although this creation and the RLDS temple might appear to be similar, when leaving the Wright creation I felt the call to aspire—and walking the famed Country Club Plaza at the church's feet, I felt the beauty of humanized space.

These are seven renderings of sacred space in one urban location, examples which I experience as having the power to draw one into contrasting worlds of meaning, and each of which expands the contours of beauty. Each is a theological World, uncovering nuances of the divine-human Spirit that can scarcely be touched in any other way. Each especially draws into itself different persons, in each case with a special space that envelops one's own pilgrimage in sacredness. Places of worship, then, are indispensable for framing and forming the human spirit.

But other needs of the human spirit need to be enveloped by sacred space. Let me illustrate two—one an imaginative domestic space, the second a public space with the power to reconcile.

Fallingwater. Frank Lloyd Wright's primal masterpiece is a home. The owner of the Pennsylvania land through which Bear Run meanders anticipated a small summer place where he and his family could enjoy a view of the modest falls. Instead, Wright built the structure *on* the falls, *in* them, *over* them, *around* them—never has "baptism" taken on such a total meaning. To be inside the space is to feel humility—in the roar of water, with rock floors sealed in such a way that they shine with what seems a perpetual wetness. At the entrance, a trickle of water into a small pool invites washing, as if in a holy water font. And a staircase in the center of the house descends down into the creek itself, as breezes flowing over the water bring cool air into the house. Cantilevering terraces jut out from various sides, so that without visible support they float over the water, much like bows of a ship, as the mist rises and the sound envelops them. The basic material is concrete, composed of materials that might have been gathered from the stream bottom itself. Parts of the exterior are faced with stone, appearing to be extensions of the cliffs; while the central fireplace is literally anchored in the central stone embankment of the site itself. It is the very stone on which the owner once liked to sit to hear the water. Even when the corner windows are opened, there is a floating lightness—for they exhibit no need of corner supports. One of my favorite dreams is that, if for only a week, this space could function as a monastery, so that the rising chants would provide Eucharistic prayers for the baptismal waters.

None of us will ever live in such a place, although we may have the gift of being able to visit. But I am convinced that beauty is not a matter of expense, as if it is dependent upon esoteric materials, or unusual sizes, or even loveliness of settings. In fact, Wright sought out problem sites for his houses, glorying in the fact that an engineer appraised the Bear Run setting as unsuited for building. Beauty is not an extra expense, as if it were something to be added—adorning a structure if one is able to afford it. The concrete he used, for example, is one of the least expensive materials, available everywhere. There is no reason why efficiency cannot be intriguing, smallness cannot be intimate, and repetition cannot have a charming rhythm. In fact, new houses selling today for upwards of half a million dollars or so are some of the ugliest structures (un)imaginable. Beauty belongs to the shape and lines and angles that define whatever space is available, and the space forms the lives of these who live within it—in nondescript ugliness or in sacredness. In the end, *beauty is the gift of imagination*, the call of creativity, the vision of life under promise, a fondness for the materials of God's creation, and a belief that the Spirit has taken up residence in human space. Indicative is the response of a couple who moved into the creation of an inspired architect: "We are not good enough for this place." Perhaps in living there, they have become so.

The Vietnam Veterans Memorial ("The Wall"). Everywhere, from the first appearance of humans, there have been memorials—remembrances, celebrations, observances, even of the dead. Their purpose is somehow to *hallow*—that is, to take something of importance and through the shaping of a particular space to render it publicly sacred. "Hallowed be thy name." Many memorials fail, for their creators fail to understand the difference between a sign as marking, and a symbol as living metaphor. A home is not a box to keep out the rain; a memorial is not a stony record of names and times. A public metaphorical creation is seldom born. What it requires is a sacredness of space that transcends what it memorializes, is far more than what remained, and holds the power to reconcile for the future those who meet there. The Vietnam Veterans Memorial ("The Wall") in Washington, DC, is such a sacred space. As a memorial, it is in truth an event—not simply to be seen, for it must be experienced. Despite its harshness, hiding nothing of human ugliness, one emerges from it changed—wherein it is beauty that is victor.

In the 1960s I traded in my "patriotism" as "idolatry" to an economic system that attempted to control violently third-world nations for its own gain. For decades I agonized for the proud nation I gloried in as a Boy Scout. I cried not only for those who died in the symbolism of that undeclared and needless war, but for those who returned— frightened, angry, feeling used, betrayed, and abandoned. They are limping crucifixes of a way of life in which we both once believed. One of my daughters took a job as an environmental lobbyist in Washington. I knew somehow I had to experience the Vietnam Wall. I asked her to experience it for both of us. This was her letter to me.

It is powerful, unlike any other I've ever seen. Most monuments glorify war and praise the brave for preserving the freedom of the nation—stuff like that. The Vietnam monument glorifies nothing. There is no white marble, no statues of heroic generals charging off to victory. A feeling of sadness permeates the whole place. Let me explain the setting. It's a wall of black stone set into the side of a gentle slope. Although it's only 200 feet from a traffic-filled street, people could pass by unaware of the monument's presence. You can't see the wall until you're right on it, just like the Vietnam War itself. The entrance is on the other side, but it too is hidden from view by a row of trees growing down the length of the Mall. It is ironic that it is most visible from the steps of the Lincoln Monument, a mass of shining white marble praising a man because of his dedication to freedom and equality. But as you approach the Vietnam Veterans Memorial, the first thing you see is a statue of three men—two white, one black. What strikes you immediately is that they are scared to death, as if in the jungle fearing for their lives, far from home, unwanted. Not exactly your typical war monument.

From there you proceed down a sloping sidewalk that runs in front of the wall. Where it starts, there are only a few names carved into the stone. But as you walk deeper and deeper, the wall gets higher and higher, and the lists of names longer and longer. In the middle, at its highest, the wall is almost 10 feet high. I got the feeling of being drawn and sucked into the monument, just as we went deeper and deeper into the war. The more blackness *there was blocking the sky—the more the killing. It is very appropriate that the wall is black. Not only that, but it is shiny, except for where the letters of the names are carved, so you see your own reflection, with the name scarred across your face. In the middle of the wall are carved a few words in dedication. Once again there is no glory, only sadness. It says something like: "This monument is dedicated to the people in the Vietnam War. The names are carved in the order that they were taken from us."*

It is moving to note the people who come to see the monument. Half of them are sightseers like me. The other half are fellow soldiers and families who have come to see the names of their loved ones who have been killed. I remember as I finally walked out along the upward slope, seeing an older couple, maybe parents, hugging each other in front of the wall, tears in their eyes. I needed to hug someone too.

I knew that I would have to go there, soon, and I did. Yet with that letter, closure was already occurring. This monument is *a table in the desert* for the pilgrimage of my country. Since then I dreamed several times of an addition to that monument. There would be another black wall, facing and paralleling the present one. On it would be carved the names of the Viet Cong. Although it would have to be far higher to accommodate the disproportionate number of dead, each wall would begin far apart, but gradually converging. Toward the middle, the passageway would become so narrow that only one person could pass through at a time. At that point the walls would intersect, so that from the center onward the sides would reverse, confusing winners and losers, victims and aggressors. The tragedy would continue to reflect back and forth between the two black walls, *ad absurdum*, until the killing halted, as if in fatigue. *Ad infinitum?* Lest we forget.

For those who still doubt, the cost of such a monument, in contrast to monuments of the past, is almost incidental. The issue is not price. The cost of beauty is inspired imagination.

—⊢—

4. The Language and Arenas of Our Space

Sacred space must not only characterize the sacred places of our explicit worship, model the meanings of our domestic spaces, and provide special public spaces for our interaction. If we take the Incarnation seriously, all of space is our concern, even to the most formative spaces of our *personal* lives. Acknowledged or not, any materials assembled around a function are a theological expression of who we are, and who, in turn, we are being shaped to be. As Winston Churchill once said, we fashion our buildings, and ever thereafter, they fashion us. But in truth what we fashion is largely, in turn, that by which we have already been fashioned. Where we worship, as we have insisted, determines in a deep sense who God is for us—and who we are becoming through relationship with that God. Thus

heretical church buildings are tragic, just as are the economic idolatries so often shaping our places of employment.
Crucial, then, *is our need to become intentional concerning all
our spaces.* More than we realize, we have the power to
choose what we will make of them—and by such choices
the meaning of Christianity for us is at stake.

Domains. The arenas of our lives are a composite of
spaces, each in need of a meaning that raises the function
involved to the level of sacramental. Blessed are those
whose multiple spaces creatively interlay and overlay and
intersect in a manner that feeds their spirits. A sign that
such awareness may be rising in our society is that, through
the use of computers, persons are beginning to find
employment that allows their working and living to share
overlapping space. Likewise, city planners are beginning to
create communities organized around campuses of green,
where work and shopping and eating and sleeping are
within walking distance, the time spent traveling between
each giving "preparation" space for entry and re-entry.

- The central arena that we have been emphasizing is
  that of the *temple* (church)—that sacred space where
  even upon solitary entrance one senses the "why" of
  one's life.
- The *synagogue* (school, education module) must
  become the space not so much for information gathering as for learning as wisdom. Here is where doing
  and being, imagining and remembering, reading and
  acting, seeing and hearing—are all given intentional

consideration so that living itself becomes understood as pilgrimage. How exciting would be a structure formed around such an understanding.

- The *home* is a spatial happening wherein a house is a domestic church—and each meal a little Eucharist. How often the lament is heard that "we rarely eat together as a family." Perhaps as an architectural response, modern floor plans often render the kitchen far more prominent than in the recent past. Instead of it being a separated space, it is made part of a "great room," so that living and cooking and eating all take place together in a common space.

A local church for which I was architectural consultant came upon their image for their new sanctuary when I noted that every time we met it was by the kitchen, complete with coffee and donuts. From that realization they easily could imagine a sacred space in which everything radiated from a common table for bread and wine. A home is where there is a "favorite" everything—each person has a "special" chair, books, tapes, pillow, color, picture, curl-up nook, a favorite tea, munchies, and a special place at the table. It is where one truly belongs, even to the point of having a space where one's cyclone habits can prevail without being forcefully inhibited. It is where solitariness and togetherness are both honored and encouraged, with a sacred "center space," functioning in its own way much as the family altar or prayer corner of old.

- The *work arena* is, in some ways, the most demeaning of contemporary spaces. Lamentable is the maze of cubicles upon cubicles, a huddling of identical cards as if dealt from a deck of vast sameness. Here are monotonous squares of isolated togetherness, of anonymous uniformity, of public privacy. In spite of such "efficiency," however, it is fascinating to observe the ingenuity of those who personalize even the space they have, adorning it ironically with pictures of mountains, rivers, and vast spaces. Even more difficult work arenas are those composed simply of open desks. Yet here too, among the transparent tape and paper clips, workers struggle to humanize their spaces most often with photographs—usually of persons with smiles, for whom their feelings are still alive.

- The *transportation module*, translated usually as "car," is for many the one remaining citadel of sanity in the midst of the chaos through which one passes daily. It is a movable hermitage, filled with the sounds of one's loving. Revealing is the ambiance—ranging from a pickup truck shouting back through open windows the decibels of country-western radio, to the respite of an alternative world of Mozartian tranquillity. Blatant bumper stickers, trophies hung from the mirror, or silvered sunglasses behind which one hides, or even a mysterious world of tinted windows. One's vehicle can be a tank of roaring aggression with up-thrust finger, or the silent luxury smugly passing all with enticing calls to envy. But alas, even here one's own space is being invaded by the cellular phone, so that nowhere can one escape being simply an extension of one's work.

But whatever and wherever the arena, deep within us is the call for a special space of one's own, marked by the imprint of one's meaning. At its best, this is not a drive to *possess* but *to establish a meaning that is genuinely mine*. And for those deprived of some such arena, the yearning will not be quieted. It simply issues forth as graffiti on every available surface.

• Perhaps most tragic symbolically is how so many persons today are deprived even of sacred space as an *arena for death's closure*. Often one takes one's last gasps in a sterile hospital, in the presence of an employee paid to be there. Even the body never "goes home," but is sent "automatically" to an artificial "space" called a funeral "parlor"—where "visitation" is reduced to a polite two-hour space. And the internment, if it is not in the totally isolated act of cremation, occurs where even the dirt is hidden by false grass. Plastic flowers mark the space of "ongoing memory." For the Church, called to the craft of spirituality, nowhere is the call more desperate than to resurrect as sacred such spatial arenas of the human spirit.

Objects. Not only is space created by being enfolded, it comes into being by the sheer presence of any "thing."

A favorite method I have to help persons get in touch with who they are is to ask them to get an object (in fact or in imagination) that they can bring back to me or a group, and with it say: "This is who I am." I still remember one woman who brought back a milkweed pod, peeled back its skin, and blew into the air the fluff that was inside. Showing the empty pod, she quietly said: "See, there's nothing inside." Each of us is fascinated by "things," and we keep them—witness the clutter of garages, attics, closets, and basements. Garage sales are little help, for they are simply the suburban game of "exchange." While our society tempts us to measure our meaning by the accumulation of possessions, the fact remains that in a certain sense we are what we collect, for what we gather around us forms us. Even we ourselves are subtle versions of our refrigerator door.

Consequently an important step in our spiritual growth occurs when we become intentional about our "things"—and thus become aware of the formative power they have in shaping and reminding, in evoking and in naming. Just as in a church building there are fonts, vestments, candles, and smells, so it is in the World of each of us. Things are the sacramentals by which the senses can be set aglow. In fact, just materials themselves are the custodians of sight and sound and smell and touch and taste. The arts of weavings, woodwork, ceramics, glass blowing, painting, incense-making, cooking—all of these beg on behalf of beauty. Through the silence of monastic meals I have discovered eating as a sacred event, my plate an orchestration of color and texture and sound and taste and smell—a veritable happening, connecting fields and sky and water and people. In a real sense, to sacralize space is to feel oneself surrounded by life: fish, pets, plants, seeds, birds, trees, animals, gardens—whatever. Then one can sense that all that is is alive, for life itself *is* the very pulsation of Spirit.

Pilgrimages. Each of us has *memory* space—of the gathered "things" of what has been. And we have a space called *present*—a composite of such images that function as foci of our emotions—in what it feels like to be who we are. For example, the space-image(s) that comes to mind when one thinks of the house where one "grew up" determines in large part the lightness or heaviness of the emotional baggage carried into the present. How different the image of a white-picket-fenced cottage, versus a door leading up to a dark attic. The present, then in turn, is integrated through the space-image(s) called the *future*—of what "might be." For some persons, this image might be the title one wants to see on one's office door, or the keys to that very special car. Yet there needs to be more. Above all we need a *meaning image*, one that distills "what it all means," in terms of the ultimate direction of one's life.

As we mentioned before, there has been throughout the Church's history a concern for pilgrimage—for a sacred space that symbolizes one's vision and goal. It is where the Christian yearns to go in order truly to "be." Classic for the Catholic might be Rome, or any of a number of shrines.

For Muslims, the yearned-for pilgrimage is likely to be Mecca, to which every Muslim who is able is obligated to make a pilgrimage at least once. For the Mormon, the place of pilgrimage may be the temple at Salt Lake City. And while the Protestant has been uneasy over any "idolatry" of place, in the past decade the "Holy Land" has laid claim to being a possible place of pilgrimage. For some people, the pilgrimage may be intensely personal—perhaps they want to return to the place of the family farm, in order to make one's peace.

Vacations are usually poor substitutes for pilgrimages. Rarely are they focused upon a place of pilgrimage *to*. Rather, they are a marathon *of*—"seeing it all" and returning home exhausted. Thus an important spiritual question for each of us, both as unique self and as Christian, is this: "Where do I need to go before I die?" I am convinced that the present fascination with labyrinths is a rebirth of this image of pilgrimage. They are not a maze, as is so much of our life. They are a path to a sacred space at the center.

Motions / Gestures. To be human is to act out meaning, and as far back as anthropologists can tell, for almost all religions this "acting out" has involved dance. In contrast, present-day Protestantism tends to restrict worship to words, and participation to seated listening. And Catholic hesitancy about experiments in dance changed only after Vatican II required that the Catholic Church in Hawaii be given special permission to allow hula as sacred dance during worship. Yet the early Church was decidedly different, as witnessed to in the phenomenon of speaking in tongues, which involves the rhythm of sound with the motion of body. Only last week I witnessed at a charismatic gathering some of the most intriguing movements. We have much to learn about spatial meaning from our youth. Somehow they are uninhibited enough to dance their way into feelings, feel their way into motion, and motion their way in special gestures all their own. Dancing in my day was a couples-thing, with dance steps intended to justify degrees of physical intimacy. Today, however, the music invites one to carve out a space on one's own. The fascination with the motion of "signing" in certain of our worship events is far deeper than courtesy to the hearing impaired. It is an expression of "acted meaning."

At this point I confess feeling deprived by my strict Protestant upbringing. Not only was any show of feeling or affection discouraged, but a number of activities such as dancing were condemned as sinful. It was not until college that I dared to try to dance. It was too late. I felt like a quasi-elephant trampling the toes of unsuspecting partners. Anything I did bore absolutely no relationship to David as he danced naked before the Ark. But in my later years I have had my moments. I remember one in particular. It was around midnight, with nothing but a full moon streaming in through the windows. I was listening to Aaron Copland's ballet, *Appalachian Spring*. For some timeless moments it happened. I dared to become lost in the moving—feeling like the soaring of a hawk, or the swaying of a cypress, and finally a sleepy cloud.

As a newly ordained priest I am finally learning the flourish of worship. Each procession is a Palm Sunday entrance. Prayerful gestures invite an indwelling of Spirit. Pushing inhibition aside with lavish bows, I have become open to the grace of kneeling and rising, and of barefooted prostration. And somehow my life all comes together as the body and blood are lifted, raised as high as I can reach—into God. The congregation also has gestures and motions. There is the touching of the forehead and lips and heart as a preparation to hear the Gospel—that one might think through the meaning, utter the response, and believe with one's whole heart. All stand at attention for the Gospel, as if to hear the "state of the union" address by the President named Jesus. Or on entering the church one's crosses oneself with water in remembrance of entering the Church at one's baptism—when one was branded and purchased at an unbelievable price.

And as we have insisted throughout, once we have been formed by such sacramentals, we can be opened to a sacredness resident in the commonplace rhythms of jogging, showering, eating, and driving. Even climbing steps can be done as sacred motion, and cooking can be done with a flourish. While I am uneasy about the analogy, I remember as a boy hearing of rich families sending their daughters to "finishing school." I have no idea what happened there, but somehow persons were to be formed in "charm." I looked up that word—it means fascination, allure, attractiveness, appeal, delight, enchantment, zest, and best of all, grace. What would it be like if the Church were able to "finish" us so that our words and actions were deeply gracious? For those folks who are still skeptical, let's try an analogy from sports: a fifty-yard pass-play, a pole vault, a double-play, a dunk—when gracefully completed, they are a choreography of beauty.

Visitations. It is interesting how often the Gospel describes Jesus in terms of *place*—that he went here, or was there. This was because wherever he went, the space became a sacred event. Likewise in the Old Testament, God's encounters with people were always *somewhere*—at a burning bush, on a volcanic mountain, in the temple, at the Jordan, within a garden. The natural response of the Hebrews was to mark sacred spaces with stones. The Church, in turn, has used architecture for marking such sites as sacred. There are no spaces of significance in Jesus' life that are not so sacralized—his birth, the Sermon on the Mount, his Crucifixion, his Resurrection, his Ascension.

Likewise, in each local Catholic Church there is a tabernacle where we can speak of divine "residency"— intentionally analogous to Israel's understanding of the Ark within the Holy of Holies. It is to such a sacred place that one is said to make a "visitation." Likewise the altar is space that God promises to visit in each Eucharist. Childhood is punctuated in meaning by "special" places— sitting on a particular limb of "my" tree, or under the porch, or by the creek, or in the closet of my room when tears were imminent. Each of us needs sacred places where there is most likely to be a "visitation" of meaning. While it is a mild form of the idea, our society points in this direction with the word "favorite." One needs a favorite restaurant, with an ambiance just right for eating with a special friend. There is a favorite path, or pond, or store, or vista, or fountain, or drive. There is a special Sunday afternoon at the zoo, or flying a kite on a special hillside with a very special grandchild. People listen dutifully on the radio to programs designed to play the "Top 40s" of the week—for even the "favorites" of others have significance.

Looking back on it now, I feel sorry for my mother. I would ask her about her favorite anything—hymn, meal, ice cream. Her answer was always the same—she replied in terms of what my father liked. To this day, I know from her all his favorites. His favorite hymn was "Faith of Our Fathers," his favorite meal was ham, and orange-pineapple was his ice cream of choice. And to this day, I do not know her favorite anything. She had none, condemned by a sexist society to have her only identity be through someone else.

—⊢—

5. From Faith to Beauty

It is not likely that one can move from experiences of beauty to a life of faith, especially in a culture where genuine beauty seems so rare. Rather, the opposite direction is most likely. For the Christian, so much depends on the divine incarnation rendering sacred the time and place of our pilgrimage. So much depends on the Holy Spirit taking up residency in the temple of one's soul. So much depends on a Christ who takes into the very heart of deity the wounds and hurts and humiliations of our lives—rendering sacred the smallest offerings of beauty, as we lift them eucharistically into the heart of God. So much depends on the sacraments, which render the Church "the house of eternal mystery, of the silent marriage of matter and spirit, nature and God." [12] When one is so claimed sacramentally, then one can accept the invitation to see everything intentionally through the eyes of faith. This does not follow easily. Yet in being formed by the Church so shaped, it can happen that for us

> *Every birth is a new creation;*
> *every greeting a prayer;*
> *every washing a baptism;*
> *every meal a Eucharist;*
> *every sleeping and waking*
> *a dying and rising in Christ—*
> *because each person*
> *is the one for whom Christ died.*

Then everything comes full circle—for then one can recognize

> *moon shadow as blue;*
> *the lemon yellow that is dawn's "break";*
> *evening's orange-red just as the birds quiet;*

and the deepest blackness called new moon;
the spring greenness of an errant shower;
the brown yellows only autumn knows; and
the white on gray of the first snow.

Christianity is not one activity among others. It is a total lifestyle of existing. Through the Church's history, monasticism has served as model for the Church's call to be a "prepared environment" of formation—a *paradisus claustralis*—a foretaste of paradise. Time after time the monks from each monastery would go into a "godforsaken" bleakness to let paradise happen again and again, for the sake of every space everywhere. Citeaux, the place heralded as the sacred space where my monastic Order began, was first a desert which the founders called "a place of horror, a vast wilderness." At a monastery one experiences a genuine alternative to the competitive materialism of present society. All is held in common, exhibiting an alternative "vision of human relationships, where beauty is more desirable than financial profit, friendship more precious than advantage, and solidarity in a common vision of human dignity more compelling than self-fulfillment." [13]

In our post-modern society, there is a sense of "a vast wilderness." And so the Church again is called to lure us beyond relativism and skepticism, in a mutual rediscovery of beauty and spirituality. The French painter Matisse understood, for he insisted that "all art worthy of the name is religious. Be it a creation of lines, or colors, if it is not religious, it does not exist." James Wall, film critic of *The Christian Century*, sees hints that we are passing into a new space. The two words he uses to signal this in present art are *ultimacy* and *intimacy*—in that we sense God in our craving for one another. My favorite words refer to the beautiful, the poetic, and the spiritual.

Thus the Church, as the residence of and the witness to sacred space as the realm of the divine presence and activity, is called to be the inner fringe at the heart of society. She is to be the earthy, visible continuation of the Incarnation, exhibiting all of space as the "theater of salvation." Her maternal role is the mission of giving birth. "What if we discover life on other planets," I've been asked, "and Jesus hadn't come there?" "No matter, I reply. God has already claimed all of space as sacred."

The Church must enter a plea for sacred spaces— both natural and created. Without deserts and wildernesses for the human soul, without the waters and oceans and rivers, the heights and depths of mountains and hills, the plains and the sunsets, the straight paths and the labyrinths, the forests and the caves—without these, the human spirit languishes, deeply. Unless our created shapes intersect with this profound need, our dwindling of spirit may be permanent. All the arts, but particularly architecture, are enterprises of formation—they tutor the heart. They enlighten how one *feels* and supplement the training of *mind* and the disciplining of *will*. A building that tells everything on the outside of what is within, is a theological failure. As with persons, the aesthetic allure is an invitation to explore interiority, one not easily emptied or domesticated. May I be forgiven for wishing that those who create ticky-tacky for the human spirit be among the first called to the heavenly judgment desk—ahead of drug dealers and even the IRS.

—⊢—

From time to time I ask myself what God's final questions of me will be. They keep changing. At this moment, I have a hunch that they may be these: Have you lived passionately, made music, given freely, prayed extravagantly, labored in prayer for the suffering, created useless beauty, attempted spontaneous kindness, pondered deeply, shaped gently what you touched—and let yourself go, deliriously, becoming lost in Me? That is, Paul, have you loved beauty?

6

THOUGHTS ON GOD

"In the face of such beauty, silence is the most appropriate response." (Anonymous Descender, 1909)

"As beautiful as the rim is, one must hike down into the inner canyon to be grasped by the spirit and mystery of this place." (Park Overlook Sign)

"The ways of the Most High have changed." (Ps. 77:10 GRAIL)

"Make a highway for the One who rides on the clouds." (Ps. 68:4 GRAIL)

One of the most brilliant philosophers of the nineteenth century, Georg W. F. Hegel, used the analogy of how our minds function for understanding God in relation to the world. For him, God thinks the world into being, which accounts for why we can understand the whole in terms of our reason. In this book, however we have been exploring the intersection of space and time as the domain where one becomes truly self-concious of who one is. We are ready now to draw the final implication of such an understanding. Becoming truly self-concious is a spiritual process, characteristic of the Christian coming of age; but, even more, it is a primal image for understanding *the very nature of* God. Rendering one's pilgrimage self-concious is the foundation of meaning. So it is with God.

Making space and time holy, then, is not only the primary human calling; it is the nature of God's pilgrimage through billions of years. Exploring the nature of God through this analogy helps resolve some major barriers to belief that have resulted from more traditional understandings. We begin with how my own pilgrimage birthed this analogy of emerging self-conciousness as a primal way of understanding both the Divine and human, and their relationship.

---

1. The Beginning: The Poet as Contemplative

No matter where one touches the sacredness of space, it ultimately confounds us until we are able somehow to come to terms with the God of space. The sacredness of space and the God of space converged powerfully for me in one of the most sacred of natural spaces—the Grand Canyon.

Hiking from the south rim of the Grand Canyon to the Colorado River floor is, physically, no more than a major challenge. The most dangerous part is not slipping after an incontinent mule team passes. Many hike it, although those who do the fourteen-mile round-trip

Kaibob Trail in one day may be a more select group. And among these, fewer still attempt the marathon journey as a spiritual pilgrimage.

My longing to do so began several years ago when, in a Navajo healing ceremony, far back in the reservation, I encountered the "Sipapu." This small hole in the hogan's dirt floor is a ceremonial focus for the power of the mysterious womb-hole from which, in Navajo belief, all spirits are birthed. The real "divine center" is known only to God, but many Native Americans believe it to be the Grand Canyon. Into that huge hole I hike, the womb from which all of us may have come. Dust to dust, life to life— on a pilgrimage from sunrise to sunset. "Seeking God" is what it turned out to be.

The first half-mile is a welcome contrast to the resort-like atmosphere at the rim. At first, my eyes become those of an *artist*. My imagination fashions a necklace from the maze of switchbacks, stringing together the mellow colors of carefree buttes and pinnacles and palisades. In time, I become more *contemplative*, empty of thought, merging and floating with the birds, now at eye level. Other persons must have been so affected, for the map identifies a point immediately in front of me as Buddha Peak, the one to the right as Vishnu Temple. Sometime after the third hour, the pilgrimage becomes decidedly *physical* and, simultaneously, more *spiritual*. Thirst is my first clue that I do not belong here. Toads, lizards, a coiled rattlesnake—they all seem at home, even a vulture overhead, circling with growing interest as my steps slow. A sign puts the matter graphically: "Danger! Those without a gallon of water each, turn back now!" Life is as thin as a canteen strap.

Panting heavily, I begin to identify with an electronic display in the museum on the rim. It is a beeper that goes off every second for three minutes. The total beeps mark the advent of the canyon within the whole span of time. As I hike, one fact becomes increasingly weighty: only with the last beep does the human species appear. In fact, human history is so minuscule that this last beep includes not only "us" but all the extinct mammoth animals of prehistory as well.

At this point I find myself encountering the experience as a theologian: philosophical by training, biblical by choice. History and tradition have become my foundation for theological exploration. But even though human figures are still vaguely visible on the rim, I have already walked beyond the symbolic equivalent of recorded history. And yet, stretching far below me, winding for miles down into the canyon, strata after strata, stretch endless symbolic layers of non-human time. Billions of years without us. Each of my steps downward is like going back in time a hundred years, as each mile-sign translates the beeps into a deepening alienation.

It is close to 10 AM when I am swept by a childlike thought. Where was God all this time? Intellectually, this is no new question. But the artistic eyes with which I began, displaced by eyes more contemplative, are now becoming very *physical* eyes, cutting straight through the romanticism of my theological metaphors. I have to get down and back out by sunset! With night temperatures well below freezing, and my water half-gone, conclusions become simple. Lost in this symbolic immensity of time, I, as a *conscious* being, am a *freak*. The whole human race simply does not belong here. Annie Dillard knows: "Either the world my mother is a monster or I am a freak."[1]

I pick up a pebble at my feet. The time it reflects, compared with the time in which self-conscious life has existed on earth, staggers the mind. In a universe possibly twenty billion years old, the first dated year in history is 4241 BC. How utterly insignificant to this bleak wholeness is the fact of self-conscious mind. Standing deep within the canyon, feeling like a humorless afterthought of a mindless whole, my operating assumption as a theologian becomes a strange *non sequitur*. How can one any longer take this recent phenomenon of self-consciousness as *the* image for understanding the meaning of the *whole*? From that point on the trail, I know myself to be a misfit, for I alone in this incredible panorama am self-conscious. Impossible to shake is a portrait of being the newest kid on the cosmic block, arrogantly insisting that behind everything is One of my own "kind."

Violating all techniques of suspenseful storytelling, let me just say straight out that I made it down and back in one day. But the price was costly. Merton claims that in the desert, one wrestles with God until one receives a new name. In the canyon that day, I know something else— that my wrestling with God shall be until God, too, is renamed. The irreversibility of this awareness becomes even clearer the next day. I drive to Mount Palomar, with its gigantic telescope. The conclusion becomes indelible. Whether I look into the earth at the Grand Canyon or away from the earth at Mount Palomar, the effect is the same. Through that telescope at this point in time, one can see one billion light-years away! This is staggering when one remembers that one light-year itself is almost six trillion miles. And while our solar system is seven billion miles in diameter, if seen from outer space it is simply one star in a gigantic galaxy which itself appears as only a minor smudge within at least 100 million observable galaxies much like our own.

What the Grand Canyon is doing for time and space, Palomar the next day would do for space and time. Self-consciousness, or even consciousness itself, is a very late arrival on a freaklike speck in an inconceivable vastness. Vastness indeed. The most distant observed galaxy thus far is in the constellation of Virgo. It is fourteen billion light-years away, with light traveling 186,281 miles a second. How, then, can one any longer propose Self-Consciousness as the defining analogy for comprehending the totality? Freud's observation can be terrifying: "I personally have a vast respect for mind, but has nature? Mind is only a little bit of nature, the rest of which seems to be able to get along very well without it."[2]

2. The Divine Autobiography

To exist is not so much to explore concepts so that one can understand them, but to search among images for *the* analogy by which to understand this tragic cosmic circus. Two generations ago, Dorothy Emmet concluded that the future of metaphysics (and thus theology) depends upon the emergence of a new analogy capable of igniting the imagination.[3] She recognized, apparently, what is becoming clearer to me now, that we have crossed a threshold in which deity as self-conscious being can no longer be entertained as preceding the cosmos. Whatever validity Christianity may claim, it must be in the full face of our cosmic loneliness, and the absurd abyss of endless aloneness for any professed deity, infinitely before any creation. The Canyon graphically illustrates the apparent *absence* of any self-conscious anything within much of the whole. Yet, as I stand on the trail, my back resting on a red bluff, I take the next mental step. As harsh as this apparent absence of God is, if we still persist in claiming a self-conscious God as the creator-designer of the whole, the question of what God has done becomes even more devastating.

It is noon now, and I am at the bottom. Beside the surging mud of the Colorado River, I eat a peanut-butter-and-jelly sandwich. The pink and orange swirl of clouds stirred the canyon into a cauldron of peach froth. Then a motion far closer refocuses my eyes. I had been watching the beauty of the sky through a spider web. And in its center is a healthy-sized spider, patiently riding the breeze, totally oblivious to the peach display. A fly strikes the web. With three venomous assaults on the terrorized insect, the spider begins sucking it apart, savoring lunch with joyous contentment.

How can I stomach a God who designed such an arrangement, especially when, sooner or later, each of us will experience the whole from the vantage point of the fly? The spider may think such an arrangement to be fine, but only until a bird sees its lunch in a spider in the middle of a web. In collecting firewood the previous day for a campfire on the rim, I saw under a log what as children we called "roly-polies." I wondered if they hurt anything. As I sit by the rapid river that endlessly tears at the canyon sides, I know the absurdity of such a question. *There is nothing alive that is not bad news for something.* From this point on, the only image of God that can any longer be viable must be one that permits me to stare bald-faced at both fly and spider.

All of this brought back to me, graphically, Ernest Becker's Pulitzer Prize–winning book, *The Denial of Death.* I am experiencing deeply his own profound wrestling with the cancer that was busily at work inside him as he wrote. Devastating it is to any God who would dare peer out from behind this death and decay woven irradicably into the fabric of "Creation," and would ask to be worshiped! Staring unblinkingly into the repulsive extravagance in time and blood that has brought evolution to where we are, Becker asks, "What are we to make of a creation in which the routine activity is for organisms to be tearing others apart…, everyone reaching out to incorporate others who are edible to him?" I was experiencing how a Creator God can no longer be held guiltless. Is God the name for that supposed "wisdom" structuring nature's bloody plan? Or is God the force that insurance underwriters actually identify as "acts of God," the One specializing in earthquakes, fires, floods, and plagues—whether willed or permitted? Or is God the One who shows up occasionally as "miracle," saving one person from an apartment house fire, while letting all the rest be torched?

In the trade we call this conundrum "theodicy," the effort to vindicate God in the face of sin and evil. Even while I lick jelly off my fingers, it becomes clear that every attempt at theodicy founders at one graphic point. Any *self-conscious* deity must be brought to trial—and condemned by only one carefully chosen picture: that of a "nightmare spectacular taking place on a planet soaked for hundreds of millions of years in the blood of all its creatures," turning the planet "into a vast pit of fertilizer."[4] The enigma that confounds every attempted theodicy rests on one beginning assumption: that from the beginning God is a fully conscious Creator. It is inconceivable that one can identify God as a loving Creator, for in Creation every organism routinely devours something else for its livelihood. As I hear a rock slide somewhere behind me, it seems graphically clear that either we have for this "terror of Creation" a sadomasochistic Designer, or an Impotent Watcher. One or the other—or else *we must forfeit self-consciousness as our informing image of God*. Saint Teresa of

Avila was enough of a mystic to put it charitably: "I do not wonder, God, that you have so few friends from the way you treat them." Similarly, Kathleen Norris once observed that God behaves in the psalms in ways he is not allowed to behave in systematic theology,

Pascal, too, drew a similar conclusion—so strongly, in fact, that he called "not true" any religion that does not hold to a God who is hidden. "Truly, you are a God who hides himself" (Isa. 45:15). In the midst of such hiddenness, I hike along the river, mesmerized by its baffling sound. I consider letting go, abandoning understanding by becoming lost in the mystery of it all. Perhaps later, but not yet. In our time, with the apparent erosion of *all* images for God, does this not condemn us to a "functional atheism"? I try to make the current "process theology" fit into this canyon experience. In the absence of any discernible pattern within history, it holds, then, that any meaning conceived as continuity must be affirmed as being within God. As a result, it is no longer "being" but "becoming" that is the ultimate category." [5] My imagination is whetted. One image that might arise is that of *space* "existing" within the consciousness of God, and matter being the visibility of God's psychic energy. Or we can identify *time* as the divine dynamic, heaven the imagination of God, hell the divine forgetting, and "kingdom" the product of God's active remembering. [6]

3. Beyond a Sad Deity

I am growing uncomfortable, with the sun vanishing early behind the deep canyon walls. I remember, in passing, the Scripture about the Son of Man having no place to put his head, at least none that is scorpion-free. Process theologians are promising, in the sense that they interpret Christian theology as having a God who does not precede Creation, but is organically related to it. God is to cosmos much as mind is to body, luring forth the novel possibilities of each instant, and taking up the redeemable into the divine Memory.[7] In the darkening shadow, I felt for this gentle, but sad, "process" God, who in the midst of "all this," somehow feels nostalgically alone. No matter how many postcard mementos of beauty are remembered, they remain what they once were—fondly remembered now.

Feeling lonely myself, I know it is time to start back. Thinking is consuming my time. I cross the bridge. That's what I need—a theological bridge, from here to I know not where—not for sure. These various images operative in Christian theology tend to offer too much or too little—either a transcendent, all-knowing God who borders on sadism, or a God so immanent within the organic process that "it" borders on impotence.

G. K. Chesterton feels right. Deeply within us, "we have to feel the universe at once as an ogre's castle to be stormed, and yet as our own cottage, to which we can return at evening." [8] Applied to the process God, I can feel pity. But this is no God against whom I can rail. This God is doing the best that is possible in a limited situation. In this image, there is no divine responsibility. My heart is beating rapidly as I reach an overlook. With such a God, the highs and the lows are leveled, diluted into a quietly reasonable and aesthetic order, kindly focused as divine invitations to choose well from divinely weighted choices. Damn. This canyon is becoming harder to climb, but somehow my soul finds an affinity here to which those riding the interstate highways are immune. The process God makes the responsibility *ours*, in which we reap from our short time what we sow, or are sown.

No. The sun is getting lower, the insects making their grand entrance. No. I refuse such a God for the price is too great. The price is a sacrifice of the full mystery, the terror, the grotesque, the hemorrhaging and blatancy of evil, as a surd worthy of rage. While their God drinks deeply of sadness, the process imagery cannot keep hidden the God of the Grand Canyon. The God here is One who dabs with blood, with a huge brush made of human hair. I would be a traitor if I denied that the cross remains the daily bottom line on history's balance sheet, inscribed in red. Annie Dillard understood this, insisting that there is a blue streak of nonbeing backed into everything that is. We are all invalids of nature, she insisted, ruminating how she has rarely seen a daddy-long-legs that has all eight legs left. The divine harshness is that the Lord made the night too long for a lot of people.

I pull out my journal and begin writing some notes. I try to figure it out on paper. If I am reading my notes correctly, an alternative is waiting to be hatched. Rosemary Ruether introduces time into the definition of God: "I am who I will *be*." [9] Carol Ochs proposes God as the place of the world.[10] Nicolas Berdyaev develops Meister Eckhart's and Jakob Boehme's mystic metaphors into "becoming"—birthing the Trinity from the Divine Nothing.[11] My mind is swirling now. What matters is if there is room for Nikos Kazantzakis's portrait of God. This God is *directly involved* in the bloody conflict of creation— as a carnal call to transubstantiate matter into divine ecstasy. His incarnate divine-human heart is "the earthen threshing-floor where night and day the defender of the borders fights with death." [12]

For a few minutes, the sunset makes thinking a preposterous enterprise. The sun makes a perfect swan dive behind the chunky horizon. These images are becoming helpful in the sense that they make it possible for me to shift our primary image of God from the *what* of self-consciousness, to the emergence itself. In fact, this makes it possible to reverse the movement in much theology. Instead of seeing movement from the transcendent to the immanent, I sense what might happen if the movement *is from immanence toward transcendence*. Somehow it seems possible to intersect two images. One is *the world as in God* ("panentheism"); the second image is *God through the world* (what we might call "panintratheism").

The Grand Canyon is a cauldron of death. I began seeing it as symbol of Creation's bloody chalice. Its restless sides teem with life, propelled by an insatiable drive to

endure, indeed, to prevail. The relentless wind, the sound-
less pull of river, the clutching root-fingers of trees, lean
varmints in crouched determination—all are sister-brothers
in this surging restlessness. We can feel this straining
deeply within ourselves. Life is thrashing about, expanding,
reaching out—in uncertain directions for seemingly
unknown reasons. But here it is that one can sense strangely
that consciousness is not totally alien. It is an amazing
breakthrough within the whole, for the sake of the whole.
Leaning out with giddiness over the rail, I could begin to
confess that this emergence of self-consciousness does
bring us to the alienating burden of knowing what nothing
else in Creation seems yet to know; and yet it opens a
deeply religious threshold. At the center of Michelangelo's
famed fresco in the Sistine Chapel appears the image of
God reaching out toward Adam, newly created. Their
fingers almost touch, as God and human finally greet in
self-consciousness, as this aperture exposes the meaning of
the Whole.

The restaurant and buildings on the rim begin to
sparkle in the crisp evening air. I feel life as an enigma—
in one sense I do not belong here beneath the edge, and
yet I have no fondness in climbing to what I will reach at
the top. Existence is polarized between the drive to be
separated *from* and the passion to lose oneself *in*. It is as if
we are characters in search of an author, for the more we
appear to "know," the more we crave to be "known"—
though not known by the hundreds of RVs around the top,
their inhabitants already abandoning thinking for a

senseless television comedy, dulling the solitude with loneliness. At the same time, I ache for a Wholeness in which the names of the dancers no longer matter. Christian mystics, those tempted to forfeit the carnal struggle, still image the birth of God in the self as paralleling God's fleshly becoming as the center point of history.[13] The primal drive to be fruitful and multiply shows an impulse to surmount the evolutionary travail beyond survival, toward participation in that which is no longer possible within a death-wrapped finitude. Thus it is that self-consciousness, evoking both terror over death and awe as its heart at worship, discloses each microcosm as the autobiography of God. And through us, God is looking out at God's own creating. And the unfathomable abyss in each of us, is the same struggle of God and nonbeing spangling the universe.

—

4. The Reverse Trinity

At this point, let me insert a few post-Canyon thoughts, triggered by the experience. Contemporary theologians have largely abandoned the once widely accepted image of a completed Creation, from which humans fell. Also discounted is the belief that at the beginning humanity was complete and whole, only subsequently having lost its wholeness as punishment for sin. The Canyon pilgrimage signals that it is time to draw the corollary concerning God. We need to abandon the image of a completed, self-conscious God, who created in completeness a cosmos that consequently alienated itself through disobedience. The dynamic of the traditional Christian understanding moves from wholeness to alienation, leaving humanity doomed to search for a reason why. The answer that began to emerge for me from the Canyon begins with less, searching for more. By the time I reached the rim, almost in darkness, I had a new primary image of God—that of a *reverse Trinity*.

The traditional understanding of the Trinity begins with the Father as a self-conscious, willing, all-powerful, all-knowing creator and designer of the whole. This God restructures the world into its present "fallen" condition as punishment for Adam's sin. Consequently, this Father sends his Son as Incarnation into the world, bridging through suffering the alienating distance universally manifested as death. Finally, God as Spirit, proceeding from the Father and the Son, comes as gift to those who believe. An official prayer for Trinity Sunday discloses the degree to which the orthodox Trinity is conceived sequentially and hierarchically, fully dependent upon the eternally self-conscious Father. "Father, you sent your Word to bring us truth and your Spirit to make us holy."[14] And it is conceived hierarchically as well, even to the point of heresy at the very beginning of every Mass—"The grace of our Lord Jesus Christ and the love of God and the fellowship of the Holy Spirit be with you all" (taken from 2 Cor. 13:13)—in which only the Father is God.

A reverse Trinity would begin with the Spirit brooding over the face of the chaos, moving from darkness out over the face of the deep, bringing form to the void. (Gen. 1–5) This sacred restlessness becomes carnal, incarnated in and through and with the length and breadth and depth of all Creation. Intertwining nature and history in a common pilgrimage, this surging Spirit we call *Holy*, for it is *Divine-self-consiousness-in-the-making*. Preconscious billions of years ago in the explosive power and riotous expansiveness of the primal cosmos, the emerging of consiousness into self-consciousness is moving powerfully toward the Divine *fullness of consciousness* whereby God is *All in all*. Expressed poetically, it is as if in the first hints of human self-consciousness millions of years ago God is gazing back over the long struggle in fond recognition, as Like hails like. All of space and time, then, is coming from, impulsed by, and returning as enrichment into the transcending Divine self-consciousness, passing into eternity as the Divine Interiority.

With this image, the "Father," as fully self-conscious Being, no longer appears at the cosmic beginning, but as the ongoing end, as the eternal culmination of the Spirit's contending. Put a bit more academically, the reversed Trinity replaces the *terminus a quo* with the *terminus ad quem*, whereby theodicy is resolved in theogony. [15] The world is God's breakthrough. Creativity is a theogonic process. We experience the world into God. And in our self-consciousness, God delights in the Creation. And above all we can experience the Holy Spirit as an unquenchable thirst for the final unity of all things gathered together as the *koinonia* or community in God. For the final consummation, Paul concludes, Christ is gathering up all things to make the final oblation to the father. And all space and time will pass into Eternity.

The new heaven and the new earth begin as hunger, bud as promise, and consummate as God's flowering in mutual self-consciousness. Thus the desert experience, so central in Christian spirituality, is, in truth, our participation in the birth pangs of God, as the Spirit, through the Son, *midwifes* the cosmos through "groaning in travail." Once back in my office from my Canyon experience, I pulled book after book from my shelf, finding hints and smidgens for clothing this image. It is really an organic reinterpretation of the social Trinity, one that entails something of the grandeur sensed by Teilhard de Chardin in the cosmic sweep toward an "Omega Point," in which everything in the universe surges toward consciousness. [16] But this divine dynamic is conceived as full drama, rather than shadow-show. The Alpha is the Spirit rather than the "Father." The Father is no longer the All-knowing Creator who navigates the whole to its predetermined harbor. This Canyon understanding is closer to Kazantzakis's God, the One bloody and panting, struggling for consciousness and thus for sanity. Here is the cosmic crucifixion, struggling toward the resurrection as fullness of being. God is immanent in all things, becoming transcendent as to self-consciousness. And Christ is "the one *coming* into the world" (John 11:27, emphasis added). God is the height, depth, breadth—and horizon.

Such thinking is in strong contrast with previous images bequeathed by science, likening the universe to a clock, with things functioning much like gears or springs. The imagery currently emerging from science is more imaginative. The ingredients of the cosmos are luminous in energy, the whole far more than the sum of all its parts, unfolding from mystery into Mystery. The new physics has restored poetry to nature. There are galaxies and clusters of galaxies and superclusters, all rotating pinwheels of billions of stars. Billions of galaxies are flung across billions of light years of infinitely expanding space—with 100,000 million stars in a galaxy, with hunches of infinite numbers of galaxies. Hugh Everett even speaks of the odds of there being an almost infinite number of universes of "other dimensions," existing simultaneously with ours, of which we can know nothing. In the end, neither space nor time make any sense of these by themselves, because space implies perimeters and time implies beginning and end— both of which "deposit" us in the Being of God.

One recalls Meister Eckhart, who claimed that our journey into the Godhead is one of spiraling through the cosmos with ever-expanding soul, making God joyful as it expands. The enormous exterior is as if one were sinking into our own center, into the inexpressible, ineffable, unnamable darkness where all becomes One, and God, he claims, is "all in all." And as if to compound Mystery upon Mystery, if the rate of expansion of the "Big Bang" had been smaller or larger by even one part in a hundred thousand million million, the universe would have either collapsed before reaching its present state, or ballooned out too rapidly for stars and planets to form, and no life would have been possible.[17]

—⊢—

5. God as a Convergence of Scriptural Imagery

This reverse Trinity is not to be found explicitly in Scripture. But neither is the traditional Trinitarian understanding. The continual task of the Church is to discern in its ongoing experiences analogies powerful enough to convert scriptural imagery into a revealing Whole. When the poet in each of us dies, God is strangled. Thus, with the image of the reverse Trinity as analogy, we can discern in Scripture an anatomy for such a God. What we can discern is God as Spirit, "crying out in every spirit," making intercessions through all of Creation "with sighs too deep for words." Prayer is that "inward groaning" in which each speck of life longs in hope to be "set free from bondage to decay," rendering "the present suffering not worth comparing with the glory to be." This becoming of God is most clear in the hungry, the thirsty, the naked, the stranger, and the prisoner, for through "the least of these" is it being "done unto me." The "deepness of our hearts" within, and "the mind of the Spirit" without, are Creation's dynamic, each searching out the other.

Evidence that "the Spirit of God really dwells in you" comes in one's heart crying out the recognition, "Abba! Father!" (Rms. 8:15) Herein we see the "plan for the fullness of time, to unite all things in [God], things in heaven and things on earth," (Eph. 1:10 RSV) so that no speck of life is any longer sojourner or stranger but is adopted into the Cosmic Household. To celebrate such a God, one can best play one of Mahler's Adagios, giving to it a libretto by the Apostle Paul. And let all of Creation join in, for the One "above all and through all and in all," whose descent as ascent is the emergence through which God is "filling all things."

Expressed Christologically, God is birthed as the lowly one, unrecognized during the hidden years, crushed and bloody in divine-human crucifixion, resurrected as the intoxicating hope of ascension luring all things as a Pentecost. The tree of Eden's temptation, planted deeply at history's tragic center, is growing as "the tree of life" whose "leaves are for the healing of the nations" (Rev. 22:10). Faith means living this vision "as if," so that one resolves the questions of theodicy through foretaste. Looking not from what once was, but back over space and time from the vantage of God's promised future, we disappear into vision. The "when" is absorbed by the "what"—for neither "mourning nor crying nor pain [shall be] any more," and thus we shall be able to say of the whole that "there shall no more be anything accursed."

—⊢—

6. Eucharist and the Divine Pilgrimage

Again, from whatever vantage we approach it, this vision is fed by the Eucharist as aperitif. Huge is the chalice of shed blood—hoisted, blessed, and returned as "the water of life without price," from which "none shall thirst again." Fermented grapes and kneaded bread are the systole and diastole of Creation's pulsing as it moves toward the Eucharistic Great Amen: "Through him, with him, in him, in the unity of the Holy Spirit, all glory and honor is yours, Almighty Father, for ever and ever. Amen!" The offertory is cosmic, as we drop the crumbs of our being and our doing into the chalice of our hopes, and lift them into God. Saint Francis of Assisi is so helpful here. He who kissed Christ incarnated in lepers, and sang to sun and moon as his sister and brother, found in the Eucharist the clue to God's infinite disguises.

This Eucharistic image of the reversed Trinity takes Hebrew Scripture seriously. It means rejecting the tendency of theologians to pick and choose from Scripture, dismissing "difficult" sections as "primitive" thinking. In contrast, we need to take Israel's antiquity earnestly—indeed not only as the emergence of human consciousness in honest travail, but also as the pilgrimage of the divine toward the fullness of Consciousness. With steady gaze down into the dereliction of the Canyon, dare we take seriously the Old Testament pictures of Yahweh's offensive extravagances, flaying with such wild ranges of feelings that the psalmists suspected a God who was as humanlike as God was divine?

Is it really faithful to Scripture to filter God into a shallow and tepid "goodness," assigning wholesale the tremors of evil to the result of human sin or to a hypothetical Satan? Far more faithful would it be to recognize this divine-human dynamic as being as internal as it is external—as the potency of Nothingness hangs uncertainly between the possibilities of creativity and of destruction.[18] Recognizable is the struggling God of plunging madness, jealously on the frantic edge of loneliness, passionately sacrificial when emerging as *agape*, disclosing the drive and passion of consciousness for its sabbatical joy. And it is in Jesus as the Christ that the same Spirit beholds itself in spirit. The Word marvels over words, in an incredible recognition of naming and being named—the God for us and the human for God.

Alfred North Whitehead names three historic transitions in our relationship to God—from God as *void*, through God as *enemy*, to God as *companion*. These he sees as marking the evolution of *human* consciousness. But if one takes the Old Testament seriously, these stages could well describe the transitions in God. It could be the way of a Wayfaring Stranger, who through the myriad yearnings and clutchings and reachings and conflicts and communions and imaginings and forgettings and rememberings, lays claim now to a proper name. Even G. K. Chesterton, custodian of orthodoxy, insisted on the Christian God as incomplete, rebellious, courageous to the breaking point, who went through agony, doubt, divine forsakenness, and was indeed, at points, an atheist.[19]

And even Isaiah dares to speak of the God who creates evil (Isa. 45:7, KJV).

Let us be clear here. We are not denying the self-consciousness of God. Rather, analogous to the primary process defining our humanness, the defining dynamic of God is that of ever increasing self-consciousness—passing beyond all that we can imagine in inclusiveness and depth, through all things "groaning in labor pains" (Rms. 8:22) into what Paul calls God's becoming "All in all." (I. Cor. 15:28) With such a vision it is possible to "comprehend, with all the saints, what is the breadth and length and height and depth. (Eph. 3:18). This incredible surge of consciousness as the Divine pilgrimage in and with and through us and all things is a relentless movement toward the *Pinacle—the Kingdom as the Divine supraconsciousness.* God is becoming transcendent through a profound immanence, becoming deeply and broadly and richly self-concious throughout the whole of creation. It is in this sense that we can experience God as what Telhard de Chardin calls the Great Presence and the Soul of souls.

God creates, purifies, and fulfills through immersion, uniting all things organically within God's own self.[20] This Divine pilgrimage, of which the Christ event is center, is thereby disclosed as the dynamic of incarnation through crucifixion into resurrection. God gathers up into God's self "our stifled ambitions, our inadequate understandings, our uncompleted or clumsy but sincere endeavors," with even the least of our desires and hopes and yearnings preserved.[21] In one sense, then, "God is everything and everything is

God, with Christ being at once God and everything."[22] The cosmic movement, of which our spirituality is a hint, is from unconciousness to preconciousness to consciousness to self-conciousness to the Divine fullness.

—⊦—

7. The Divine Sexuality

The interaction of nature and spirit into consciousness finds its most powerful imagery, perhaps, in *sexuality*. While sexual acts are sometimes ruthless in their drivenness, as in certain animals, for humans sexuality can become a sacrament without parallel. In fact, Paul uses sexual imagery for the relationship of Christ and his Church. Creation is an androgynous love affair of the Spirit with the Lamb (Rev. 22:17), consummated as the marriage feast that celebrates the emerging God as All in all. Sexuality, spanning foreplay to afterbirth, wraps richly this analogy for God—writhing, gentle, wrenching, joyous, desperate, ecstatic, embracing, frightening, exciting, lonely, clutching, loving, jealous, laughing, bloody, teasing, agonizing, frantic, unrepeatable. Understandably, Becker insists that God is the only adequate sex partner.

Nor is it little wonder that Christianity has been obsessed with sexuality through the centuries. Desirous of controlling sexuality as a frightening and mysterious drive, the Church has never permitted sexuality to serve as a divine image. This chaotic interplay of sexual joy and childbirth pain as analogy for the Whole would be a conundrum—with emerging images of violence conflicting with an interface of infinite gentleness. Sexuality can emerge with terrifying violence, and yet it is rooted, Donald Bloesch insists, in the passionate human yearnings of the soul for God.[23] This may be why Christians keep stumbling over issues such as birth control and abortion. Down deep, such struggles evoke a strange feeling—that our attempted parochialism is a sad effort to domesticate the Wild Spirit, until the free eagle comes to resemble a pigeon.

I am back home now, in my monastic hermitage. In this quiet cedar forest I find that I am unable to let go of the Grand Canyon experience—and my caustic struggle with God and evil. But on this crisp fall morning, I awoke with a thin remembrance of another experience, this on a winter day, here at the hermitage. Somehow I sensed that these two experiences might have been made for each other. I spent the morning searching my journals for a description of it. This is the way my journal records it.

> This morning I have had the God-experience for which I have yearned for so long. I now know what it means to name the Name—for to experience God means to smell the flames of the fireplace behind me, to see the sun straining to break the haze haloing the cedars in front, and to hear the straining and zinging and cracking of the frozen lake. Everything has the expectancy of a Cezanne still life, waiting the red dash of a cardinal's arrival. I turned on a tape of Victoria's "O Magnum Mysterium."

The ache of my soul and the yearning of all things around me are finding a friend in each other. I am breathless with "knowing." It is no longer an Absent Landlord with whom I am dealing—One who, before vacationing in parts unknown, structured a world with mindless clumsiness, and put out a "Do Not Disturb" sign on his shop door. God is the Spirit pulsating within, and incarnated everything without. Everywhere this dawn, I see and hear the body of God's writhing and ecstasy. God pulses in my mind and loins—even in my knees, aching to trudge the snow, to roll and frolic and burrow.

Yet I dare never blink away the realism. It is deathly cold outside, fifteen below zero, with a wind that sears the flesh. Animals have begun their daily death-hunt for the careless and the unfortunate. Yet it feels different now. There is no puzzle behind the mystery, for the Mystery itself is sufficient. All is here in front of me—and in me—as liturgical dance.

In being released from a lifelong guessing game, I am free to experience the divine recognition—in a child's giggle, a soaring hawk, the pleading of eyes—and, can I say it?, in the screaming terror of a rabbit, crunched into unconsciousness by a hungry coyote. Can such a God be conscious? Yes, at the edge of becoming—sometimes barely, other times brilliantly, slightly ahead of each threshold. Is God present? Everywhere, enormous in breadth, expansive in depth, and beyond us all in imagination and memory. God is the emerging consciousness which darts in and out, through and for, behind and in

front, to be encountered as Self-Consciousness in reve-
latory moments. God is the place of the world, history
the continuo of God's time—and faith a wagering on
God's determination to emerge with us and to outlast the
ongoing threat of Nothingness.

Was it Merton who likened humankind to Jacob, wrestling with a gutsy God for a blessing, on the way home to an estranged twin? I see now that the Grand Canyon and the Cedar Forest are two sides of the same reality. Theodicy is of little consequence unless lived. And what is emerging is a spirituality of the reverse Trinity—centering in the Triune God as Source, Companion, and Goal. Creation, Redemption, Consummation. As the Creative Source, God the Spirit is the grounding and hungry abyss of being—as midwife. As Incarnate Companion, God the Son/Daughter is the relationality and mutuality of Being—as wedded lovers and beloved friends. And as self-conscious goal, God the Father/Mother is the ecstatic fullness of being—coming, coming, ever coming. And through it all, our subsistence is in God. All things are in God as Transcendence; God is in all things as Immanence. "Show us thy face," pleads the psalmist—over and over again. In Christ, God does just that. The face is kindly, gentle, forgiving, smiling, caring, compassionate—and deeply in love. And in that sense, Christ is the only adequate image of God.

How is one to live this emerging image? It means responding to God as the inner and outer edge of everything, yearning from the inside out, and the outside in—at one's fingertips, where breath touches lungs, where smell connects flower—with one's soul being God's birthing within. We are the emerging soul of the universe, called as poets to be midwives for the soul in everything. Spiritual experience, then, ranges from mystic homesickness, through the intimacy and contending of new lovers and old friends, to the communion and consummation of one's soul, lost in the One named Love. Or in two words, this is the God of the *Cosmic Triduum*.[24] And what matters is that when my life ends, I shall have made some small contribution to the becoming of God. And the goal of that becoming? The creation of beauty, Whitehead suggests.

I am the spiritual director for a person declared to be the most violent prisoner in the entire federal prison system. Unable to be controlled in the highest security prison, a specially built cell was constructed, where he has been in solitary confinement for seventeen years. Only once has his cell been entered—when one morning, shackled and handcuffed, surrounded by armed guards, he received the water of Baptism. I knew it was for real when from the tiny feeding slot in his steel-plated door I could see leaves from a watermelon seed he had "planted" with a wet napkin in a Dixie cup.

7

CONCLUSION

"In you all find their home." (Ps. 87:7 GRAIL*)*

"My happiness lies in you alone." (Ps. 16:2 GRAIL*)*

"Let everything that breathes praise the LORD*." (Ps. 150:6)*

I remember as a boy sitting one night with my dog, staring at the stars. The dog refused to look. I focused his eyes by holding his nose in the right direction. Every effort met with failure. And when finally I realized that the stars could not exist for him, a shadow fell between us. Self-consciousness rendered me, and all of us, lonely aliens in the Grand Canyon of our world. And it terrorizes us as well, for it brings with it an awareness of the eons of violence that have been expended in birthing our emergence. Reconciliation with such a process is impossible for me, or for the human race, if our *self-consciousness* is the *imago Dei* by which we conceive God as preceding the whole as its designer. Reconciliation is possible, however, if the *imago Dei* is identified as the *creative emergence* of self-consciousness. It is an *event* rather than a *thing*. Consciousness, then, is the surging, inchoate interiority of all things, emerging as a transcendence who greets our aloneness as friend and ally.

What binds divine and human metaphysically is the ache and passion for wholeness, pleading for recognition in the Other. And faith? It is a commitment to see the world with an analogy so illuminating that in choosing it, the experience becomes one of having been chosen.

On the night of my Grand Canyon event, I had dinner at a restaurant on the rim. It was quite some time before I realized why that was necessary. My table was at a window, where outside a full moon teased the canyon into playful shadows. Most of the people in the restaurant wore suits and gowns—and I had not even cleaned my sneakers. A woman at the next table was making a vocation out of humiliating the waiter—first by demanding a glass without water spots, followed by a fork that wasn't bent, and a napkin that needed to be better laundered. She and the others around her were oblivious to what that huge black hole right outside their window could do to them—permanently. I confess that even with a meal framed by freshly baked bread and vintage wine, I still did not fully understand. It wasn't clear until much later why on that evening I toasted with bread and wine the Grand Canyon outside my window.

Only after watching many mornings as the valley beneath my monastery was made a chalice of frothy fogwine, only then did I realize that the Grand Canyon had become my *chalice*. Even more, *the Grand Canyon and the Triduum are actually two sides of one reality*. What the Triduum is for the Church, the Grand Canyon is in the world. One is the rhythm of desert liturgy, of going down in order to come up. The other is the Sacrament of sacraments, the Eucharist, of raising up in order to receive back. Together they transform our eyes, our ears, our touch, even our taste and smell, so that all of space and all of time are sacralized in foretaste. *Alleluia. Alleluia. Alleluia.*

What can it mean to *be* Eucharist? I mostly have images. It must have something to do with being *taken, broken, blessed,* and *given.* Above all, there comes the image of the early Church as a community of feasting: they found in eating and drinking the fulfillment of Christ's promise of Resurrected Presence. And this Presence consumed is the incarnation continued in each of us. While the Eucharist is the source and summit of the Christian life, it is as well the empowering rhythm that informs the cosmos and dances with it into its meaning. For the world, Eucharist provides the recipe for manna, and thus the menu for those who do not know what to do with life's ingredients. We are not dealing with works of merit, but a gift of grace, for in each Eucharist of each day, the death and resurrection of Jesus Christ is made sacramentally present and lively. The Incarnation continues when, as the liturgy says, through the mystery of water and wine we share in the divinity of Christ, who humbled himself to share in our humanity. Equally powerful is the image of the walking wounded, shuffling forward at each Eucharist to receive—as a host is lifted before their eyes, with the promise: "The Body of Christ." Indelible is the image I have of lifting the face of a dying cancer victim from his own vomit and anointing his parched lips with the blood of Christ—as *viaticum*, as food for the journey. There is the image of standing at the altar, lifting skyward as far as one can reach, the paten of broken lives and the chalice of poured-out pain. This is the moment of knowing, *truly* knowing, that despite all theology and every doctrine, what alone matters is the ongoing "transubstantiation" of time and space into the Crucified God—to be received back as an empowerment of the Risen Christ. Living eucharistically is to become *a* gift, through *the* Gift. It is to *be* Eucharist.

If someday, God forbid, I find that somehow my mind is no longer able to affirm Real Presence, I pray that I shall still never stop the offering. That which keeps the world from collapsing is the "handful" of those who are willing, in our different ways, to raise the crucifixions of the world into the "being" of God's, and, in turn, to put God into the mouth of those whose suffering is a birthing at soul-depth.

—⊢—

We have explored *space* as the glorious arena of God's becoming, filling as well every nook and cranny of our own daily lives with the call to creativity and the sheer joy of being alive. But in the end, to be able to live this depends for the Christian on the richness of the Church—within which we stand in order to see. On the one hand, we need to stand there with H. Richard Niebuhr, praying for forgiveness for the present state of the church-world of multiple denominations. "The road to unity is the road of repentance."[1] Especially is this the case when "denomina-tionalism" renders "doctrine" a hypocritical excuse for division unfaithfully originated, needlessly fed, and per-mitted out of habit.

And yet, on the other hand, our differences *can be* the rich facets by which the jewel of the Church can be seen as foretaste of the divine glory. Before we had any such thing as "denominations" in the modern sense of the word, the Church had an incredible number of religious orders, expressing with a host of theological and ethical accents the rich fabric of Christian space. This panorama ranges still, from the Benedictines (sixth century) to the Augustinians (thirteenth century), the Carthusians (eleventh century) to the Dominicans (thirteenth century), the Franciscans (thirteenth century) to the Jesuits (six-teenth century). And Saint Bernard of Clairvaux, looking out upon this rich tapestry of his own day, shared words equally valid as a perspective for understanding the sacred space of our own day.

> I admire them all. I belong to one of them by obser-vance, but to all of them by charity. We all need one another. The spiritual good which I do not own and possess, I receive from others.... In this exile, the church is still on pilgrimage and is, in a certain sense, plural. She is a single plurality and a plural unity. All our diversities, which make manifest the rightness of God's gifts, will continue to exist in the one house of the Father, which has many rooms.

Now there is a division of graces; then there will be distinctions of glory. Unity, both here and there, consists in one and the same charity.

APPENDICES

*"I remember the days of old.... Teach me
the way I should go." (Ps. 143:5, 8)*

*The following appendices offer some practices that readers may
find helpful as a means of making their space holy.*

DAILY PRAYERS

Two short prayers for use during the first moments after awakening and as the last thing before sleep-as brackets for making holy the space of one's daily life.

Morning Prayer

Yours is the day, O God, and yours is the night.
You have established the sun and the moon.
The lands of sunrise and sunset, you fill with joy.
It is you who have fixed the boundaries of the earth.
Summer and winter you have made.
Parent of the orphan, lover of the widow,
You give the lonely a home, and lead prisoners to freedom.
-May I walk with you this day? Amen.

Evening Prayer

You are in our midst, O Lord, your name we bear.
Do not forsake us, O Yahweh our God.
Into your hands, Lord, we commend our spirits.
If I live, I live to the Lord; if I die, I die to the Lord.
So whether I live or whether I die, I am the Lord's.
Into your hands, Lord, I commend my spirit.
Keep me as the apple of your eye,
Protect me in the shadow of your wings,

And cradle me in your everlasting arms.
Into your hands, Lord, I commend my spirit.
—May the Almighty God grant us a restful night
And a peaceful death. Amen.

JOURNALING: AN INTRODUCTION

Journaling is a way of bringing self-consciousness into the process of living, so that one can see how the space and time of one's life form a sacred pilgrimage.

There are probably as many ways of journaling as there are persons who do it. One's method usually develops over time as one becomes increasingly clear as to why one is doing it. Journaling, most often, begins as variations on several themes: (1) to "deal with" and "drain away" negative and potentially damaging feelings; (2) to provide a surrogate "friend" with whom one might explore one's deeper self; (3) to seek the meaning of one's life by searching for an overarching "design," in the light of which one's life is rendered a "pilgrimage."

These needs are present within all of us, but in varying degrees of intensity and personal investment, depending on the particular time of our lives in which we are journaling. Elaborate journaling techniques are available (e.g., Ira Progoff workshops), but in almost all cases one ends up creating different sections within the journal. Most persons will have a section for making an "unthinking" recording of "data" (e.g., "Today seemed to be nothing but an endless series of meetings. It began when…"). Soon one will likely want to record or later extract elements to be placed in easily identified sections of the journal (e.g., "Perennial

Problems," or "Recurring Themes," or "Illuminations"). By creating such sections, one becomes involved in what I call a "theologizing" process. This is similar to how and why ancient Israel told stories of the events of their people around the campfire. In the process, certain stories began to be favorites, incorporating the other stories. This was how folks became aware of more overarching themes. These stories, in turn, would be seen as simply different ways of pointing to *the* story, which would be so identified from its continual repetition.

We call "redactors" those persons who pulled together the various stories, laws, and prophecies from Israel's telling and retelling, forming them into a whole. They did this by discerning among the whole a story that presented the primal meaning that the other material echoed. Thus, for example, if one wants to understand the context in which the Israelites functioned, one reads the story of Genesis 1, where God created the heavens and the earth. And if one wants to understand the human condition as this came to be recognized through Israel's long struggles, one reads chapters 2 and 3 of Genesis; about a garden and a tree. And if one wants to understand God's ongoing response to Israel's waywardness, one reads the story of the exodus (e.g. Exodus 15) It's all there in these stories—so powerfully that the rest of the Scripture is visible as variations on these primal themes.

So is the case with the Gospel writers. Jesus taught primarily through telling stories, invoking rich imagery that in turn evoked meanings beyond the words. He told a multitude of stories, but only a limited number were remembered and finally recorded—because they were the ones that spoke most powerfully to the particular Gospel writer's own dilemma, in need of healing. Thus, to understand who Jesus was personally for Luke, we need to become immersed in the story of the Prodigal Son. Luke keeps telling us variations of this story, such as those of the lost sheep and the mislaid coin, presumably because it spoke to his own need so powerfully.

This background helps us understand the journaling process, which is much the same: recording and gathering together the events, feelings, patterns, and themes of one's life. As with Israel, a major goal of journaling is to discern increasingly in the "many" the underlying theme of the "all."

In my own journaling, I use dividers. The largest section is for recording "*Entries.*" I try not to put off writing these for more than two or three days—otherwise freshness quickly fades into inane generalities. My effort is to preserve what seems important, without any need at the moment to know why. Then I have another section labeled "*Past.*" This section has come to have subsections, as I experience hints of meaningful life divisions. They may have external markings, such as "Early Childhood" and "High School Years." For some persons, a better division may be according to major events, either positive or negative—such as "After My Parent's Divorce" or "Before My Sister Was Born." The purpose of such divisions is to get in touch with "who I really am," and, correlatively, "why I tend to do what I do." I enter there incidents or statements that pop into my mind, often as evoked by daily entries. Another section is marked "*Insights*." Here my entries might be about myself, other persons, or the state of the world. I have learned that even the great "eternal truths," to which I occasionally come, have a life-expectancy of less than an hour—unless recorded.

Another section I entitle "*Possibilities*." Progoff has a similar section which he calls "Roads Not Taken." Here a past event invites future possibilities. For me, this section has the flavor of "If only I would have…" or "Is it possible that…." Another section I have found useful is labeled "*Unfinished Agenda*." This can contain things as simple as a painful letter that needs to be answered, to a relationship festering because of inactivity. While I have great difficulty remembering dreams, many persons profit greatly from a section marked "*Dreams*"—in which re-occurrences and themes can be observed and detected.

My final section is called "*Life Themes*," which is my primary theologizing section. It is really a "section about sections." It is a "final" place of distilling, reserved for special insights or breakthroughs. Most often these are variations on the themes of regrets, concerns, and hopes.

Whatever sections or subsections a person comes to create, I suggest that they begin with one's daily entries and then emerge out of need rather than conforming to a preconceived structure. Otherwise the beginner will be confronted (as with such sections as I am using) with a complexity that is inhibiting. Simply begin by creating sections as opportunities to store "data," until one discovers repetitions, patterns, or even gross accumulations that warrant special distillation and reflection.

A variation on this journaling approach is to use *time* as one's primary structure. One can use the same approach as described above for recording data. Then one sets aside a regular time, perhaps an afternoon near the end of each month, for special journaling. Here one reads and re-reads the daily entries for the month, recording in a special section one's discernment of "what has really been happening to me this month." Throughout the year, one does this for each month in turn. Toward the end of the calendar year, most meaningfully on New Year's Eve or New Year's Day, one reads through the entries in the section containing the monthly discernments for the year. The task here is to do for the whole year the kind of discerning one has been doing for each month, asking: "What has this year really been for me?" Here one permits to flow the thanksgivings, emerging opportunities, perennial negativities, etc., and records them in a section marked "Years."

When I use this process, I then read the summary entries that are gathered in the "Years" section. In doing this, what I find amazing, in one sense, is how much of significance *has* occurred. And yet, in another sense, I am amazed how very *little* has occurred—and I experience a deep gnawing in my soul as I read the same negative entries, year after year. Then can emerge a "constructive guilt," in which I am driven to set as a priority making some change in my life. This journaling method, then, is a way of "distilling time," so that one can begin to identify a still-point around which all else in one's life takes its meaning.

It is also possible to journal creatively around the theme of *space*. Again one writes the daily entries as already indicated. Then one creates sections around the "places" of one's living—at first the most obvious will be spaces such as home, work, and/or school. Here one becomes increasingly attentive not only to the "what" of the events of one's life, but to the feel of the "space" in which they occur—positively and negatively, hurtfully and imaginatively. Before long one will find emerging those special arenas that give one's life its particular flavor—as joys and aches and longings. For many of us, such aware-ness leads in two important directions. One can come to identify where, unawares, one actually is being fed. This can lead to a powerful awareness of one's need for places in order "to be": a "desert" for silence, an area to "unwind," a place to be "inspired," an environment in which to "re-energize." Such journaling may also raise flags of differing colors—about where one's vitality is being squeezed or emptied out, where dis-ease and ugliness is threatening and eroding one's meaning. *A Table in the Desert* can serve well as a text with which one can journal, rendering it personal as an inventory and resource.

Throughout, it is important never to lose perspective on *why* one is journaling. It is an instrument for discerning *wholes* through *particulars*. If this is forgotten, the danger is that the process will become an end in itself ("keeping" a journal). Reduced to a tiresome duty, one will soon stop.

Or even if one persists, the journal will function more as an escape than an engagement.

One last suggestion: Journals must be absolutely confidential, except where one might choose to share a particular entry with one's spiritual director. If one plans to do this often, a loose-leaf binder is helpful, with a line down the middle of the page—the left side for you, the right side for the director. But in no case should one journal with *someone else* in mind—even the director. If one does, one will almost always filter what one writes. Further, one's journaling will have minimal usefulness if one has even a faint suspicion that it may not be totally confidential. John Wesley developed a code for his journal, so that even if someone found it, no one could read his entries. But, alas, a scholar recently has broken the code. Now poor John's interior life with all its temptations is on public exhibition. If you cannot find a safe place for your journal, preferably where it can be locked, label it with the most innocuous title that you can imagine. Mine is labeled "Lectures in Metaphysics." Guess I will now have to change the title.

appendix | 3

LECTIO DIVINA (SACRED READING)

Lectio Divina means "sacred reading." It is a method practiced throughout Christian history whereby Scripture becomes a sacred vehicle through which God speaks uniquely to each soul. This method for providing illumination and empowerment by the Spirit can be practiced alone or in groups.

Making space and time holy requires spiritual methods by which these dimensions can be discerned. Of all the techniques that the monastic tradition has explored and bequeathed us, perhaps *lectio divina* is the most important and most widely used. It is a slow, meditational approach to the reading of spiritual books, especially Scripture. Monks often "pray" the psalms by singing them—hearing with "the ear of our hearts." At other times, especially in the darkness of night, they do "reverential listening" in the silence. *Lectio divina* stands in significant contrast to "speed reading." Instead of skimming for content, one savors every word. In fact, the word "meditation" literally means "tasting with one's lips." Thus the ancient monks took it for granted that reading Scripture meant reading it out loud, chewing it, as it were. A helpful analogy is how the impact of poetry is multiplied when read aloud, especially by the poet.

The important first step in *"lectio"* is choosing the passage of Scripture. It might be a favorite, or it might simply be the passage that appears next as one works chronologically through the Old or New Testaments, or both. Just as Mary "treasured all these words and pondered them in her heart" (Luke 2:19), so this is what the Christian is to do. Read and re-read the passage, which might be only a verse or two, until a phrase or a word attracts one. Mull it over, taste it, feel the sound of it, let oneself be grasped by it. Consider if it could be God's special word for your day— and then treat it as such. One may need to read a longer passage, and re-read it several times, until the attraction occurs.

The second stage is "*oratio*," meaning prayer as *conversation*. Here one permits the "word" to stir in the ashes of one's memory, or within the imagination of one's openness. Share these as a communication with God, as one might with a special friend. "Ask, and it will be given you; search, and you will find; knock, and the door will be opened for you" (Luke 11:9).

Then one permits one's special word or phrase for the day to become a mantra, repeating it so that it functions much like a toy to keep the mind out of the way. Gradually one fades into a profound sense of Presence, much as an older couple can sit for hours by the fireside, speaking not a word, but being deeply rooted in each other's contentment. This step, called *contemplation*, is a matter of re-falling in love, forgetting oneself by resting in the Other. Fr. Luke Dysinger, OSB, suggests that *lectio divina* teaches us to savor and delight in all the different flavors of God's presence.

This method has also been used well in group settings. The group chooses a section of Scripture and then reads through it quite slowly. After a period of silence, the passage is read another time. Out of the silence, each person is invited to share the special word or phrase that emerges for them. During a final reading, one is to mediate upon that word in the context in which it appears, imagining the implications if incarnated into one's own life. These discernments are then shared—not in any full sense, but in terms of a pregnant image or possibility. When finished, each member silently uses that word or image as a mantra, as together the group sinks into the Presence of the One. However used, *lectio divina* is a process by which one is graced with a gift from the past, discerned in the present, for the future as promise.

A SAMPLE LIST OF WHAT TO REMEMBER DAILY

"Making holy" is integrally related to the need to remember. Christians tend to be faithful in what they recall they have promised to do in order to be a serious Christian. But during the in-between times, unless one keeps a disciplined guard against one's natural forgetfulness, our reversion to secular life is rapid. This "Sample List" is a model by which one can begin to create positive habits. Success in the spiritual life comes when through discipline it becomes habitual.

A Daily Motto:

"You, God, are all I have, and you are all I need; my life is in your hands."

1. **Ponder**

   Permit yourself to think with the Spirit about "the meaning of life"—about the big and deep things, about life and death—focusing on the God who is deeply within and vastly without.

2. **Practice the Presence**

   Live in intentional awareness of God as immanent companion. Develop a relationship through which, in your own small way, you might help bring God increasingly into Being through your own deepening consciousness.

3. Contemplate

Set aside a special time in which through silence you can become lost in God, in foretaste of the end-time.

4. Intercede

Lift particular persons and concrete things into God—developing a willingness, if need be, to take their place in their dilemma and/or suffering.

5. Image

Center your day by trying to make each action a Eucharistic gesture—offering it up to God, where it is blessed and returned as a healing to be shared. Rehearse and facilitate this posture by frequent participation in the Eucharist itself.

6. Anchor Points

Establish special times, such as at the hourly beep of your watch or when passing a particular picture or spot in your home, for remembering that you are unconditionally loved by God—now, always.

7. Spontaneity

Schedule regularly a day in which you arise without having your space and time planned, content to discern as the day unfolds what God's concrete yearning for you might be.

8. Let God

On a regular basis, enjoy a Sabbath Day—by refraining from doing anything that "has" to be done, or "feels like work"—being thankfully open to the varied gifts that come simply from "being."

9. Simplicity

Exercise the discipline of rendering your lifestyle increasingly simple—in dress, food, etc. Surround yourself with uncluttered space: keep only possessions that you presently use, and use nonplastic materials that celebrate the natural beauty of stone and wood and clay and fabric.

10. Hospitality

Practice spontaneous giving with no thought of return, and surprise with beauty those persons who would least expect it.

11. Co-creation

In your own special way, be a co-creator with God— enjoying creativity for its own sake. Dare to become involved in writing poetry, sculpting, weaving, painting, gardening, building—and breathing, deeply.

12. Confession

Make a full confession of your past, ideally to another person who can keep confidences. One may do this through journaling—or at least continue to do so over time, adding and expanding and deepening in increasing honesty about one's life. By confessing, begin to see how you might atone for the hurts that can in some sense be undone, and let forgiveness suffice for those that cannot be undone. And be clear that when true forgiveness occurs, one should never again mention that which through forgiveness no longer exists. So, through confession, by being freed of the past, and rendered less needy about the future, strive to remain in the fullness of the present.

13. Sanctification

Understand your pilgrimage as a "growth in grace." Thus through either individual or group spiritual direction, receive enough support and accountability that you lose your taste for the temptations by which secular society attempts to control us. These can be summarized as prestige (earning acceptance), possessions (craving security), and seeking power (using others). Be clear that this striving for purity in motivation is fed by the "purity of heart" to will one thing—that of seeking God for its own sake. Play some Mozart just to be sure.

14. Living As If

One's love affair with God and with God's creation involves praying in the full face of the world's agony. Permit yourself to be pained by the suffering that lurks as close as one's neighbor, and is as vast as the whole earth. Vow to live now that radical life which Christ promised as characterizing the age to come. If the Kingdom is to be nonviolent, so should one's life be totally nonviolent now. And if the coming of Christ brings that space in which there shall be no more hunger or thirst, so shall I live my life to help make this possible now—"for the former things are passing away."

A SAMPLE RULE

"A Sample List of What to Remember Daily" is important as a beginning discipline. But as one lives it experimentally, one should begin constructing a Rule that one might sign, promising to be held faithful to it. What follows might serve as a sample.

A "Rule" is a concrete promise, usually in writing, to which a person or group is willing to be held accountable for living life in a specific manner, one that is as meaningful as one can realistically imagine. The Church has long regarded a "rule" as the base for the discipline that is indispensable for serious growth in grace. Each religious "Order" and many Protestant denominations have a corporate "Rule." The following sample might help a person sense those areas that are worth being included in one's personal rule. The more concrete one's rule, the more useful it will be.

I Shall:
1. Engage in daily prayer, especially upon rising and before retiring. And I shall "labor in prayer" for those persons in need of my intercessions.
2. Read the Bible daily, possibly using a lectionary to suggest passages.
3. Participate regularly in corporate worship, and celebrate the Eucharist as often as it is available.

4. Read at least one book a month that will expand and deepen my knowledge and faith-style as a committed Christian.

5. Keep abreast of the daily news—locally, nationally, and worldwide—from several perspectives. I shall do this with a special attention to what I can do— in person, through my resources, and by encouraging corporate responsibility, as in seeking the passage of certain legislation.

6. Care for my body as a temple of the Spirit—by means of a healthy diet, regular exercise, and prudence in using stimulants and drugs.

7. Endeavor to establish a creative balance between work and leisure; seriousness and play; engagement and silence; family/ friends and myself—with openness to rhythms that help establish wholeness.

8. Engage in a periodic examination of the space of my faith-style: including my house, its furnishings, appliances, and tools; my personal possessions and accumulations; my work environment and especially the spaces there that tempt me to gossip, etc.; variety, simplicity, and purity of recreation; the ecology of my transportation. I will be intentional about the goals of simplicity, low consumption, energy efficiency, shared possessions, and responsible stewardship.

9. Take several hours monthly to appraise how my money is being spent, how my time is used, and what spaces center my life. I will strive intentionally

to make my actions and my professed beliefs a consistent whole. In order to gain both perspective and centering, I vow to take periodic retreats away from my normal activities and spaces, in such a place as a monastery or retreat center.

10. Intersect my life with life—such as with plants, animals, trees, gardens, parks, a zoo. I will do what I can to redeem the livability of the environment—locally, nationally, and internationally.

11. Move toward a tithe (10 percent) of my income—to be shared by the Church and social justice ministries that are concerned for persons and for the humanizing of socioeconomic structures.

12. Intentionally live my life according to holy time, cognizant of making more meaningful such segments as days, weeks, years, etc.

13. Intentionally form my life by being surrounded by holy space—with an eye to simple and earthy beauty, resisting the ornate, the artificial, and the commercial. This means being energized by form, shape, line, and rhythm, perceiving these in nature and creating them in the spaces of my living. I will extend such sacredness with whatever talents I have, creating useless beauty with my hands and spontaneous compassion in my relating.

14. Find a friend, spiritual director, or group that can provide the supportive accountability I need in order to be faithful to this rule.

EUCHARISTIC SCRIPTURAL REFERENCES

The following are the primary New Testament references to the Eucharist (Lord's Supper), listed in the order in which they were probably written. These are supplied to encourage one to ponder the importance of the Eucharistic act as central to one's living.

> 1 Corinthians 11:17-34
> Mark 14:22-25
> Luke 22:14-20
> Matthew 26:26-29
> John 6:47-58

Other references to the Eucharist:

> Luke 24:30-35
> John 6:23-35
> Acts 1:13; 2:42; 2:46-47; 20:7; 20:11; 27:35
> 1 Corinthians 5:7-8; 10:1-5; 10:16-22; 12:12-13
> Hebrews 7:11; 13:10-13
> 1 John 5:6-8
> Jude 12
> Revelation 19:9

ENDNOTES

chapter two: space-time

1. St. Benedict, *Rule for Monasteries*, (Collegeville, Minn., 1935), chapter 31.
2. Joseph Martos, *Doors to the Sacred* (Liguori, Mo.: Triumph Books, 1981), 127-8.
3. Gerard Manley Hopkins, "God's Grandeur," in *The Poems of Gerard Manley Hopkins* (New York: Oxford University Press, 1975), 31.
4. *Book of Blessings*, National Conference of Catholic Bishops (Collegeville, Minn.: Liturgical Press, 1989).
5. Edward M. Hays, *Prayers for the Domestic Church* (Easton, KS: Shantivanam House of Prayer, 1979).
6. Dorothee Soelle, *Death by Bread Alone* (Philadelphia: Fortress, 1978), 132-36.
7. *Voyage, Vision, Venture*, Task Force on Spiritual Development of the American Association of Theological Schools in Theological Education (Spring 1972), 23-24.
8. Robert M. Hamma, *Landscape of the Soul: A Spirituality of Space.* (Notre Dame, IN: Ave Maria Press, 1999).
9. Soelle, *Death by Bread Alone*, see pp. 69, 102, 134, 147.

chapter three: the world of sacrament

1. John Calvin, *Institutes of the Christian Religion*, vol. 2 (London: James Clarke and Co., 1953), 289.
2. Ernst Troelsch, *The Social Teachings of the Christian Church*, vol. 2 (New York: Macmillan, 1931), 993f.

3. See Mary Margaret Funk, *Thoughts Matter* (NY: Continuum, 1998).

4. See W. Paul Jones, *A Season in the Desert: Making Time Holy* (Brewster, Mass.: Paraclete Press, 2000). Chapter 7 gives a full description of the Triduum.

5. Oscar Cullman, *Essays on the Lord's Supper* (Richmond: John Knox, n.d.), 12.

6. Quoted in Harold E. Fey, *The Lord's Supper* (New York: Harper and Row, 1948), 18.

7. Louis Evely, *The Church and the Sacraments* (Denville, OH: Dimensions Press, 1971), 39.

8. *The United Methodist Book of Worship* (Nashville: The United Methodist Publishing House, 1992), 43.

9. Kathleen Norris, *The Cloister Walk* (New York: Riverhead Books, 1996), 266.

10. Thomas Merton, *New Seeds of Contemplation* (New York: New Directions, 1962), 290-97.

chapter four: sacraments and sacramentals

1. For a description of the Triduum see Jones, *A Season in the Desert*, Chapter 7.

2. Frank C. Senn, ed., *Protestant Spiritual Traditions* (New York: Paulist Press, 1986), 71.

3. *The Book of Discipline of the United Methodist Church* (Nashville: The United Methodist Publishing House, 1996), 61-62.

4. *Handbook for Today's Catholic* (Liguori, MO: Liguori Publications, 1994), p. 46, with references to the

Catechism of the Catholic Church (Washington: United States Catholic Conference, 1994), sections 1382–1398; 1402; 1405.

5. *Handbook for Today's Catholic* (Liguori, Mo.: Liguori Publications, 1994), 91.

chapter five: the shaping of space as beauty

1. Quoted in "Vita Consecrata," in *Origins*, vol. 25, no. 41 (April 4, 1996).

2. Gerard Manley Hopkins, "God's Grandeur," in *The Poems of Gerard Manley Hopkins*, 66.

3. Andre Louf, *Tuning into Grace* (Kalamazoo, Mich.: Cistercian Publications, 1992), see pp. 65, 67.

4. Cistercian Breviary (Trappist, Ky.: Gethsemani, 1961), 116-17.

5. The hymn *"Ternis Ter Horis Numerus,"* for the office of None during Lent, written before the eleventh century; translated at Gethsemani Abbey, 1970.

6. *Lenten and Easter Seasons*, vol. 2 of *The Liturgy of the Hours* (New York: Catholic Book Publishing Co., 1976), 496-98.

7. Hopkins, *The Poems* , 66.

8. See Jones, *A Season in the Desert*.

9. Jerry Filteau, "Architect Criticizes Church Design Trend, Blames Guidelines," Catholic News Service, September 1999.

10. See W. Paul Jones, *Theological Worlds: Understanding the Alternative Rhythms of Christian Belief* (Nashville:

Abingdon, 1989); W. Paul Jones, *Worlds Within a Congregation: Dealing with Theological Diversity* (Nashville: Abingdon, 2000).

11. Filteau, "Architect Criticizes Church Design."

12. Introduction to *Theologia Germanica* (New York: Pantheon, 1949), 53.

13. Words of Paul Philibert as quoted by Kathleeen Norris, *The Cloister Walk* (New York: Riverhead Books, 1996), 378.

chapter six: thoughts on god

1. Annie Dillard, *Pilgrim on Tinker Creek* (New York: Bantam Books, 1974), 180.

2. Sigmund Freud, *Psychoanalysis and Faith* (New York: Basic Books, 1963), 133-34.

3. Dorothy Emmet, *The Nature of Metaphysical Thinking* (London: Macmillan and Company, 1945).

4. Ernest Becker, *The Denial of Death* (New York: The Free Press, 1973), 282-85.

5. Philosophically, the impetus was evident in Charles Hartshorne, *The Divine Relativity* (New Haven: Yale University Press, 1948); theologically, the first significant notice was John B. Cobb Jr., *A Christian Natural Theology* (Philadelphia: Westminster Press, 1965).

6. E.g., Marjorie Suchocki, *God, Church, World* (New York: Crossroads Press, 1982) and *The End of Evil: Process Eschatology in Historical Context* (Albany, N.Y.: State University of New York Press, 1988); Eugene Peters, *The*

Creative Advance (St. Louis: Bethany, 1966); David R. Griffin, *Process Christology* (Philadelphia: Westminster Press, 1973).

7. John B. Cobb Jr., *Process Theology as Political Theology* (Philadelphia: Westminster Press, 1982) and with David Ray Griffin, *Process Theology: An Introductory Exposition* (Philadelphia: Westminster, 1976).

8. G. K. Chesterton, *Orthodoxy* (New York: Dodd, Mead and Co., 1950), 130.

9. Rosemary Radford Ruether, *Sexism and God-talk: Toward a Feminist Theology* (Boston: Beacon, 1983).

10. Carol Ochs, *Behind the Sex of God* (Boston: Beacon, 1977).

11. Nicolas Berdyaev, *The Destiny of Man* (London: Geoffrey Bles, 1948), 29f.

12. Nikos Kazantzakis, *The Saviors of God* (New York: Simon and Schuster, 1960), 151 *et passim*.

13. E.g., Origen, St. Gregory of Nyssa, Evagrius of Pontus, Richard of St. Victor, Julian of Norwich, and St. Teresa of Avila.

14. Opening prayer for Trinity Sunday, *The Sacramentary* (New York: Catholic Book Publishing Co., 1974), 346.

15. E.g., Leonard Hodgson, *The Doctrine of the Trinity* (New York: Charles Scribner's Sons, 1944); Charles Lowry, *The Trinity and Christian Devotion* (New York: Harper, 1946); C. C. J. Webb, *God and Personality* (London: G. Allen and Unwin, 1918); Lionel Thornton, *The Incarnate Lord* (London: Longmans, Green, 1928).

16. Teilhard de Chardin, *The Future of Man* (New York: Harper and Row, 1964); *The Phenomenon of Man* (New York: Harper, 1959).

17. See *Christian Century*, June 2–9, 1999, and *The Christian Science Monitor*, March 11, 1999.

18. The distinction between "to ouk" (nothingness) and "to meon" (nonbeing) is useful in indicating the two contrary dimensions involved here.

19. G. K. Chesterton, *Orthodoxy*, 138-39.

20. Teilhard de Chardin, *The Phenomenon of Man* (New York: Harper, 1959), 293.

21. Teilhard de Chardin, *Hymn of the Universe* (New York: Harper, 1969), 120, 139.

22. *Ibid*, 125.

23. Donald Bloesch, *The Struggle of Prayer* (San Francisco: Harper and Row, 1980).

24. See Jones, *A Season in the Desert*, chapter 7

chapter seven: conclusion

1. H. Richard Niebuhr, *The Social Sources of Denominationalism* (Hamden, Conn.: Shoe String Press, 1929), 284.